MEMOIRS
of the
American Mathematical Society

Number 948

Yang-Mills Connections on Orientable and Nonorientable Surfaces

Nan-Kuo Ho
Chiu-Chu Melissa Liu

[redacted]

November 2009 • Volume 202 • Number 948 (second of 5 numbers) • ISSN 0065-9266

American Mathematical Society
Providence, Rhode Island

2000 *Mathematics Subject Classification.*
Primary 53D20; Secondary 58E15.

Library of Congress Cataloging-in-Publication Data

Ho, Nan-Kuo, 1975-
Yang-Mills connections on orientable and nonorientable surfaces / Nan-Kuo Ho, Chiu-Chu Melissa Liu.
p. cm. — (Memoirs of the American Mathematical Society, ISSN 0065-9266 ; no. 948)
"Volume 202, number 948 (second of 5 numbers)."
Includes bibliographical references.
ISBN 978-0-8218-4491-5 (alk. paper)
1. Geometry, Differential. 2. Moduli theory. 3. Yang-Mills theory. I. Liu, Chiu-Chu Melissa, 1974- II. Title.

QA641.H65 2009
516.3′6—dc22 2009029177

Memoirs of the American Mathematical Society

This journal is devoted entirely to research in pure and applied mathematics.

Subscription information. The 2009 subscription begins with volume 197 and consists of six mailings, each containing one or more numbers. Subscription prices for 2009 are US\$709 list, US\$567 institutional member. A late charge of 10% of the subscription price will be imposed on orders received from nonmembers after January 1 of the subscription year. Subscribers outside the United States and India must pay a postage surcharge of US\$65; subscribers in India must pay a postage surcharge of US\$95. Expedited delivery to destinations in North America US\$57; elsewhere US\$160. Each number may be ordered separately; *please specify number* when ordering an individual number. For prices and titles of recently released numbers, see the New Publications sections of the *Notices of the American Mathematical Society.*

Back number information. For back issues see the *AMS Catalog of Publications.*

Subscriptions and orders should be addressed to the American Mathematical Society, P. O. Box 845904, Boston, MA 02284-5904 USA. *All orders must be accompanied by payment.* Other correspondence should be addressed to 201 Charles Street, Providence, RI 02904-2294 USA.

Memoirs of the American Mathematical Society (ISSN 0065-9266) is published bimonthly (each volume consisting usually of more than one number) by the American Mathematical Society at 201 Charles Street, Providence, RI 02904-2294 USA. Periodicals postage paid at Providence, RI. Postmaster: Send address changes to Memoirs, American Mathematical Society, 201 Charles Street, Providence, RI 02904-2294 USA.

This publication is indexed in *Science Citation Index*®, *SciSearch*®, *Research Alert*®, *CompuMath Citation Index*®, *Current Contents*®*/Physical, Chemical & Earth Sciences.*
Printed in the United States of America.

∞ The paper used in this book is acid-free and falls within the guidelines established to ensure permanence and durability.
Visit the AMS home page at http://www.ams.org/

10 9 8 7 6 5 4 3 2 1 14 13 12 11 10 09

Contents

Abstract

In "*The Yang-Mills equations over Riemann surfaces*", Atiyah and Bott studied Yang-Mills functional over a Riemann surface from the point of view of Morse theory. In "*Yang-Mills Connections on Nonorientable Surfaces*", we study Yang-Mills functional on the space of connections on a principal $G_{\mathbb{R}}$-bundle over a closed, connected, nonorientable surface, where $G_{\mathbb{R}}$ is any compact connected Lie group. In this monograph, we generalize the discussion in "*The Yang-Mills equations over Riemann surfaces*" and "*Yang-Mills Connections on Nonorientable Surfaces*". We obtain explicit descriptions of equivariant Morse stratification of Yang-Mills functional on orientable and nonorientable surfaces for non-unitary classical groups $SO(n)$ and $Sp(n)$. When the surface is orientable, we use Laumon and Rapoport's method in "*The Langlands lemma and the betti numbers of stacks of G-bundle on a curve*" to invert the Atiyah-Bott recursion relation, and write down explicit formulas of rational equivariant Poincaré series of the semistable stratum of the space of holomorphic structures on a principal $SO(n,\mathbb{C})$-bundle or a principal $Sp(n,\mathbb{C})$-bundle.

Received by the editor July 23, 2007, and in revised form September 28, 2007.

Article electronically published on July 22, 2009; S 0065-9266(09)00564-X.

2000 *Mathematics Subject Classification.* Primary 53D20; Secondary 58E15.

Key words and phrases. Moduli space, Yang-Mills connections, Morse stratification.

The first author was supported by Grant NSC 95-2115-M-006-012-MY2 and NSERC Postdoctoral Fellowship.

Acknowledgments

We thank Tom Baird, Ralph Cohen, Robert Friedman, Paul Goerss, Lisa Jeffrey, Eckhard Meinrenken, John Morgan, David Nadler, Daniel Ramras, Paul Selick, Michael Thaddeus, Jonathan Weitsman, and Christopher Woodward for helpful conversations. We thank Gérard Laumon and Michael Rapoport for confirming our understanding of their paper [**LR**].

CHAPTER 1

Introduction

Let $G_{\mathbb{R}}$ be a compact, connected Lie group. The complexification G of $G_{\mathbb{R}}$ is a connected reductive algebraic group over $\mathbb{C}$. For example, when $G_{\mathbb{R}} = U(n)$, then $G = GL(n, \mathbb{C})$. Let P be a C^∞ principal $G_{\mathbb{R}}$-bundle over a Riemann surface Σ, and let $\xi_0 = P \times_{G_{\mathbb{R}}} G$ be the associated C^∞ principal G-bundle. The space $\mathcal{A}(P)$ of $G_{\mathbb{R}}$-connections on P is isomorphic to the space $\mathcal{C}(\xi_0)$ of $(0,1)$-connections ($\bar{\partial}$ operators) on ξ_0 as infinite dimensional complex affine spaces. In the seminal paper [**AB**], Atiyah and Bott obtained results on the topology of the moduli space $\mathcal{M}(\xi_0)$ of (S-equivalence classes of) semi-stable holomorphic structures on ξ_0 by studying the Morse theory of the Yang-Mills functional on $\mathcal{A}(P)$. The absolute minimum of Yang-Mills functional is achieved by *central* Yang-Mills connections, and $\mathcal{M}(\xi_0)$ can be identified with the moduli space of gauge equivalence classes of central Yang-Mills connections on P. When the absolute minimum of the Yang-Mills functional is zero, which happens exactly when the obstruction class $o(P) \in H^2(\Sigma, \pi_1(G)) \cong \pi_1(G)$ is torsion, the central Yang-Mills connections are flat connections, and $\mathcal{M}(\xi_0)$ can be identified with the moduli space of gauge equivalence classes of flat connections on P.

Atiyah and Bott provided an algorithm of computing the equivariant Poincaré series $P_t^{\mathcal{G}^{\mathbb{C}}}(\mathcal{C}_{ss};\mathbb{Q})$, where $\mathcal{C}_{ss}$ is the semi-stable stratum in $\mathcal{C}(\xi_0)$ and $\mathcal{G}^{\mathbb{C}} = \mathrm{Aut}(\xi_0)$ is the gauge group. They proved that the stratification of $\mathcal{C}(\xi_0)$ is $\mathcal{G}^{\mathbb{C}}$-equivariantly perfect, so

$$P_t^{\mathcal{G}^{\mathbb{C}}}(\mathcal{C}(\xi_0);\mathbb{Q}) = P_t^{\mathcal{G}^{\mathbb{C}}}(\mathcal{C}_{ss};\mathbb{Q}) + \sum_{\lambda\in\Xi'_{\xi_0}} t^{2d_\mu} P_t^{\mathcal{G}^{\mathbb{C}}}(\mathcal{C}_\mu;\mathbb{Q})$$

where d_μ is the complex codimension of the stratum $\mathcal{C}_\mu$, which is a complex submanifold of $\mathcal{C}(\xi_0)$, and the sum is over all strata except for the top one $\mathcal{C}_{ss}$. The left hand side can be identified with $P_t(B\mathcal{G};\mathbb{Q})$, the rational Poincaré series of the classifying space $B\mathcal{G}$ of the gauge group $\mathcal{G} = \mathrm{Aut}(P)$. On the right hand side, $P_t^{\mathcal{G}^{\mathbb{C}}}(\mathcal{C}_\mu;\mathbb{Q})$ can be related to the equivariant Poincaré series of the top stratum of the space of connections on a principal G_μ-bundle, where G_μ is a subgroup of G. So once $P_t(B\mathcal{G};\mathbb{Q})$ is computed, $P_t^{\mathcal{G}^{\mathbb{C}}}(\mathcal{C}_{ss};\mathbb{Q})$ can be computed recursively. Zagier solved the recursion relation for $G = GL(n,\mathbb{C})$ in [**Za**], and Laumon and Rapoport solved the recursion relation for a general connected reductive algebraic group G over $\mathbb{C}$ in [**LR**]. The series $P_t^{\mathcal{G}^{\mathbb{C}}}(\mathcal{C}_{ss};\mathbb{Q})$ can be identified with $P_t^{G_{\mathbb{R}}}(V_{ss}(P);\mathbb{Q})$, where $V_{ss}(P)$ is the representation variety of central Yang-Mills connections on P. When the obstruction class $o_2(P) \in H^2(\Sigma;\pi_1(G)) \cong \pi_1(G)$ is torsion, $V_{ss}(P)$ is the representation variety of flat Yang-Mills connections on P, which is a connected component of $\mathrm{Hom}(\pi_1(\Sigma), G_{\mathbb{R}})$.

In [**HL4**], we study Yang-Mills functional on the space of connections on a principal $G_{\mathbb{R}}$-bundle P over a closed, connected, nonorientable surface Σ. By pulling back connections to the orientable double cover $\pi : \tilde{\Sigma} \to \Sigma$, one gets an inclusion $\mathcal{A}(P) \hookrightarrow \mathcal{A}(\tilde{P})$ from the space of connections on P to the space of connections on $\tilde{P}$, where $\tilde{P} = \pi^* P \to \tilde{\Sigma}$. The Yang-Mills functional on $\mathcal{A}(P)$ is the restriction of the Yang-Mills functional on $\mathcal{A}(\tilde{P})$. For nonorientable surfaces, the absolute minimum of the Yang-Mills functional is zero for any P, achieved by flat connections. The moduli space of gauge equivalence classes of flat $G_{\mathbb{R}}$-connections on P can be identified with a connected component of $\mathrm{Hom}(\pi_1(\Sigma), G_{\mathbb{R}})/G_{\mathbb{R}}$, where $G_{\mathbb{R}}$ acts by conjugation.

In this paper, we generalize the discussion in [**AB**] and [**HL4**] in the following directions:

(1) In Chapter 2, we compute the rational Poincaré series $P_t(B\mathcal{G}; \mathbb{Q})$ of the classifying space of the gauge group $\mathcal{G}$ of a principal $G_{\mathbb{R}}$-bundle over any closed connected (orientable or nonorientable) surface. The case where Σ is orientable is known (see [**AB**, Theorem 2.15], [**LR**, Theorem 3.3]).
(2) When Σ is orientable and $G_{ss} = [G, G]$ is not simply connected (for example, when $G = G_{ss} = SO(n, \mathbb{C})$, $n > 2$), the recursion relation [**LR**, Theorem 3.2] that Laumon and Rapoport solved in [**LR**] is not exactly the Atiyah-Bott recursion relation [**AB**, Theorem 10.10]. As a result, their formula for $P_t^{ss}(G, \nu'_G)$ [**LR**, Theorem 3.4] is not exactly $P_t^{\mathcal{G}^{\mathbb{C}}}(\mathcal{C}_{ss}(\xi_0); \mathbb{Q})$ when G_{ss} is not simply connected. In Appendix A, we show that the method in [**LR**] inverts the Atiyah-Bott recursion relation and yields a closed formula for $P_t^{\mathcal{G}^{\mathbb{C}}}(\mathcal{C}_{ss}(\xi_0); \mathbb{Q}) = P_t^{G_{\mathbb{R}}}(V_{ss}(P); \mathbb{Q})$, where $G_{\mathbb{R}}$ is any compact connected real Lie group (Theorem 4.4, Theorem A.9).
(3) In [**HL4**], we established an exact correspondence between the gauge equivalence classes of Yang-Mills $G_{\mathbb{R}}$-connections on Σ and conjugacy classes of representations $\Gamma_{\mathbb{R}}(\Sigma) \to G_{\mathbb{R}}$, where $\Gamma_{\mathbb{R}}(\Sigma)$ is the super central extension of $\pi_1(\Sigma)$. This correspondence allows us to obtain explicit description of $\mathcal{G}$-equivariant Morse stratification by studying the corresponding representation variety of Yang-Mills connections. In Chapter 4, we recover the description in terms of Atiyah-Bott points for orientable Σ, and determine candidates of Atiyah-Bott points for nonorientable Σ.
(4) In Chapter 5, Chapter 6, and Chapter 7, we give explicit descriptions of $\mathcal{G}$-equivariant Morse strata of Yang-Mills functional on orientable and nonorientable surfaces for non-unitary classical groups $SO(2n+1)$, $SO(2n)$, and $Sp(n)$. When Σ is nonorientable, some twisted representation varieties (introduced and studied in Section 4.6 and Section 4.7) arise in the reduction of these non-unitary classical groups. This is new: in the $U(n)$ case (see [**HL4**, Section 6, 7]), the reduction involves only representation varieties of $U(m)$, where $m < n$, of the nonorientable surface and of its double cover.
(5) When Σ is orientable, we use the closed formula in (2) to write down explicit formulas for $P_t^{G_{\mathbb{R}}}(V_{ss}(P); \mathbb{Q})$ for non-unitary classical groups (Theorem 5.5, Theorem 6.4, and Theorem 7.4). These formulas are analogues of Zagier's formula for $U(n)$.

The topology of $\mathrm{Hom}(\pi_1(\Sigma), G_{\mathbb{R}})/G_{\mathbb{R}}$ is largely unknown when Σ is nonorientable. Using algebraic topology methods, T. Baird computed the $SU(2)$-equivariant cohomology of $\mathrm{Hom}(\pi_1(\Sigma), SU(2))$ and the ordinary cohomology of the quotient space $\mathrm{Hom}(\pi_1(\Sigma), SU(2))/SU(2)$ for any closed nonorientable surface Σ [**B**]. He also proposed conjectures for general G.

For the purpose of Morse theory we should consider the Sobolev space of L^2_{k-1} connections $\mathcal{A}(P)^{k-1}$ and the group of L^2_k gauge transformations $\mathcal{G}(P)^k$ and $\mathcal{G}^{\mathbb{C}}(P)^k$, where $k \geq 2$. We will not emphasize the regularity issues through out the paper, but refer the reader to [**AB**, Section 14] and [**Da**] for details.

CHAPTER 2

Topology of Gauge Group

Let Σ be a closed connected surface. By classification of surfaces, Σ is homeomorphic to a Riemann surface of genus $\ell \geq 0$ if it is orientable, and Σ is homeomorphic to the connected sum of $m > 0$ copies of $\mathbb{RP}^2$ if it is nonorientable.

Let $G_{\mathbb{R}}$ be a compact connected Lie group. Let P be a principal $G_{\mathbb{R}}$-bundle over Σ, and let $\mathrm{Aut}(P) = \mathcal{G}(P)$ be the gauge group. When Σ is orientable, the rational Poincaré series $P_t(B\mathcal{G}(P);\mathbb{Q})$ was computed in [**AB**, Section 2] for $G_{\mathbb{R}} = U(n)$. The computation can be generalized to any general compact connected Lie group (see [**LR**, Theorem 3.3]). In this section, we will compute $P_t(B\mathcal{G}(P);\mathbb{Q})$ when $G_{\mathbb{R}}$ is any compact connected Lie group and Σ is any closed connected (orientable or non-orientable) surface.

Following the strategy in [**AB**, Section 2], we first find the rational homotopy type of the classifying space $BG_{\mathbb{R}}$ of $G_{\mathbb{R}}$ (see [**Se**]). Note that $BG_{\mathbb{R}}$ is homotopic to BG, where G is the complexification of $G_{\mathbb{R}}$. Let $H_{\mathbb{R}}$ be a maximal torus of $G_{\mathbb{R}}$. Then $H_{\mathbb{R}} \cong U(1)^n$, and

$$H^*(BH_{\mathbb{R}};\mathbb{Z}) \cong \mathbb{Z}[u_1,\ldots,u_n],$$

where $u_i \in H^2(BH_{\mathbb{R}};\mathbb{Z})$. The Weyl group W acts on $H^*(BH_{\mathbb{R}};\mathbb{Q}) \cong \mathbb{Q}[u_1,\ldots,u_n]$, and

$$H^*(BG_{\mathbb{R}};\mathbb{Q}) \cong H^*(BH_{\mathbb{R}};\mathbb{Q})^W \cong \mathbb{Q}[I_1,\ldots,I_n]$$

where I_k is a homogeneous polynomial of degree d_k in $u_1,\ldots,u_n$. We may take $I_k \in \mathbb{Z}[u_1,\ldots,u_n]$, so that $I_k \in H^{2d_k}(BG_{\mathbb{R}};\mathbb{Z})$. We may assume that $d_1 = \cdots = d_r = 1$, and $d_k > 1$ for $k > r$. Then $r = \dim_{\mathbb{R}}(Z(G_{\mathbb{R}}))$, where $Z(G_{\mathbb{R}})$ is the center of $G_{\mathbb{R}}$. In particular, $r = 0$ if and only if $G_{\mathbb{R}}$ is semisimple. The classes $I_1,\ldots,I_n$ are the universal characteristic classes of principal $G_{\mathbb{R}}$-bundles. Each $I_k \in H^{2d_k}(BG_{\mathbb{R}};\mathbb{Z})$ induces a continuous map $I_k^* : BG_{\mathbb{R}} \to K(\mathbb{Z};2d_k)$ to an Eilenberg-MacLane space, so we have a continuous map

$$BG_{\mathbb{R}} \to \prod_{k=1}^{n} K(\mathbb{Z}, 2d_k).$$

This is a rational homotopy equivalence.

Fact 2.1. Let $\stackrel{\mathbb{Q}}{\simeq}$ denote rational homotopy equivalence. Then

$$BG_{\mathbb{R}} \stackrel{\mathbb{Q}}{\simeq} \prod_{k=1}^{n} K(\mathbb{Z}, 2d_k)$$

In addition to Fact 2.1, we need the following two results:

Proposition 2.2 ([**AB**, Proposition 2.4]).

$$B\mathcal{G}(P) \simeq \mathrm{Map}_P(\Sigma, BG_{\mathbb{R}}),$$

where the subscript P denotes the component of a map of Σ into $BG_{\mathbb{R}}$ which induces P.

THEOREM 2.3 (Thom).

$$\operatorname{Map}(X, K(A,n)) = \prod_q K(H^q(X,A), n-q)$$

where $K(A,n)$ is the Eilenberg-MacLane space characterized by

$$\pi_q(K(A,n)) = \begin{cases} 0 & q \neq n \\ A & q = n \end{cases}$$

Since $\pi_q(X \times Y) = \pi_q(X) \times \pi_q(Y)$, we have

$$K(A_1 \times A_2, n) = K(A_1, n) \times K(A_2, n).$$

Let Σ be a Riemann surface of genus ℓ. Then

$$\begin{aligned}
\operatorname{Map}\Big(\Sigma, \prod_{k=1}^n K(\mathbb{Z}, 2d_k)\Big) &= \prod_{k=1}^n \operatorname{Map}(\Sigma, K(\mathbb{Z}, 2d_k)) \\
= & \prod_{k=1}^n \Big(K(H^2(\Sigma;\mathbb{Z}), 2d_k - 2) \times K(H^1(\Sigma;\mathbb{Z}), 2d_k - 1) \times K(H^0(\Sigma,\mathbb{Z}), 2d_k)\Big) \\
= & \Big(\mathbb{Z} \times K(\mathbb{Z},1)^{2\ell} \times K(\mathbb{Z},2)\Big)^r \\
& \times \prod_{k=r+1}^n \Big(K(\mathbb{Z}, 2d_k - 2) \times K(\mathbb{Z}, 2d_k - 1)^{2\ell} \times K(\mathbb{Z}, 2d_k)\Big)
\end{aligned}$$

where the factor $\mathbb{Z}^r$ corresponds to different connected components. So

$$\begin{aligned}
\operatorname{Map}_P(\Sigma, BG_{\mathbb{R}}) \overset{\mathbb{Q}}{\simeq} & \Big(K(\mathbb{Z},1)^{2\ell} \times K(\mathbb{Z},2)\Big)^r \\
& \times \prod_{k=r+1}^n \Big(K(\mathbb{Z}, 2d_k - 2) \times K(\mathbb{Z}, 2d_k - 1)^{2\ell} \times K(\mathbb{Z}, 2d_k)\Big).
\end{aligned}$$

It follows that

THEOREM 2.4 ([**LR**, Theorem 3.3]). *Let $B\mathcal{G}$ be the classifying space of the gauge group $\mathcal{G}$ of a principal $G_{\mathbb{R}}$-bundle over a Riemann surface of genus ℓ. Then*

$$P_t(B\mathcal{G}; \mathbb{Q}) = \left(\frac{(1+t)^{2\ell}}{1-t^2}\right)^r \prod_{k=r+1}^n \frac{(1+t^{2d_k - 1})^{2\ell}}{(1-t^{2d_k - 2})(1-t^{2d_k})}.$$

Note that $P_t(B\mathcal{G}; \mathbb{Q})$ does not depend on the topological type of the underlying principal $G_{\mathbb{R}}$-bundle.

Let Σ be the connected sum of $m > 0$ copies of $\mathbb{RP}^2$. Then

$$\mathrm{Map}\Big(\Sigma, \prod_{k=1}^{n} K(\mathbb{Z}, 2d_k)\Big) = \prod_{k=1}^{n} \mathrm{Map}(\Sigma, K(\mathbb{Z}, 2d_k))$$

$$= \prod_{k=1}^{n} \Big(K(H^2(\Sigma;\mathbb{Z}), 2d_k - 2) \times K(H^1(\Sigma;\mathbb{Z}), 2d_k - 1) \times K(H^0(\Sigma,\mathbb{Z}), 2d_k)\Big)$$

$$= \prod_{k=1}^{r} \Big(\mathbb{Z}/2\mathbb{Z} \times K(\mathbb{Z}, 1)^{m-1} \times K(\mathbb{Z}, 2)\Big)$$

$$\times \prod_{k=r+1}^{n} \Big(K(\mathbb{Z}/2\mathbb{Z}, 2d_k - 2) \times K(\mathbb{Z}, 2d_k - 1)^{m-1} \times K(\mathbb{Z}, 2d_k)\Big)$$

where the factor $(\mathbb{Z}/2\mathbb{Z})^r$ corresponds to different connected components. So

$$\mathrm{Map}_P(\Sigma, BG_{\mathbb{R}}) \overset{\mathbb{Q}}{\simeq} \prod_{k=1}^{n} \Big(K(\mathbb{Z}, 2d_k - 1)^{m-1} \times K(\mathbb{Z}, 2d_k)\Big)$$

It follows that

THEOREM 2.5. *Let $B\mathcal{G}$ be the classifying space of the gauge group $\mathcal{G}$ of a principal $G_{\mathbb{R}}$-bundle over a non-orientable surface which is diffeomorphic to the connected sum of $m > 0$ copies of $\mathbb{RP}^2$. Then*

$$P_t(B\mathcal{G}; \mathbb{Q}) = \prod_{k=1}^{n} \frac{(1 + t^{2d_k - 1})^{m-1}}{(1 - t^{2d_k})}.$$

For classical groups we have:

(A) $G_{\mathbb{R}} = U(n)$: $W \cong S(n)$, the symmetric group, so

$$H^*(BU(n); \mathbb{Q}) = \mathbb{Q}[u_1, \ldots, u_n]^{S(n)} = \mathbb{Q}[c_1, \ldots, c_n],$$

where c_k is the k-th elementary symmetric function in $u_1, \ldots, u_n$. In fact, the generator $c_k \in H^{2k}(BU(n); \mathbb{Q})$ is the universal rational k-th Chern class. So $d_k = k$, $k = 1, \ldots, n$.

(B) $G_{\mathbb{R}} = SO(2n + 1)$: $W = G(n)$, the wreath product of $\mathbb{Z}/2\mathbb{Z}$ by $S(n)$, so

$$H^*(BSO(2n + 1); \mathbb{Q}) = \mathbb{Q}[u_1, \ldots, u_n]^{G(n)} = \mathbb{Q}[p_1, \ldots, p_n],$$

where p_k is the k-th elementary symmetric function in $u_1^2, \ldots, u_n^2$. In fact, $p_k \in H^{4k}(BU(n); \mathbb{Q})$ is the universal rational k-th Pontrjagin class. So $d_k = 2k$, $k = 1, \ldots, n$.

(C) $G_{\mathbb{R}} = Sp(n)$: $W = G(n)$, the wreath product of $\mathbb{Z}/2\mathbb{Z}$ by $S(n)$, so

$$H^*(BSp(n); \mathbb{Q}) = \mathbb{Q}[u_1, \ldots, u_n]^{G(n)} = \mathbb{Q}[\sigma_1, \ldots, \sigma_n],$$

where σ_k is the k-th elementary symmetric function in $u_1^2, \ldots, u_n^2$. So $d_k = 2k$, $k = 1, \ldots, n$.

(D) $G_{\mathbb{R}} = SO(2n)$: $W = SG(n)$, the subgroup of $G(n)$ consisting of even permutations, so

$$H^*(BSO(2n); \mathbb{Q}) = \mathbb{Q}[u_1, \ldots, u_n]^{SG(n)} = \mathbb{Q}[p_1, \ldots, p_{n-1}, e],$$

where p_k is the k-th elementary symmetric function in $u_1^2, \ldots, u_n^2$, and $e = u_1 \cdots u_n$. In fact, $p_k \in H^{4k}(BU(n); \mathbb{Q})$ is the universal rational k-th

Pontrjagin class, and $e \in H^{2n}(BSO(2n); \mathbb{Q})$ is the universal rational Euler class. So $d_k = 2k$, $k = 1, \ldots, n-1$, and $d_n = n$.

CHAPTER 3

Holomorphic Principal Bundles over Riemann Surfaces

Let G be the complexification of a compact, connected real Lie group $G_{\mathbb{R}}$. Then G is a reductive algebraic group over $\mathbb{C}$. For example, if $G_{\mathbb{R}} = U(n)$ then $G = GL(n, \mathbb{C})$. We fix a topological principal $G_{\mathbb{R}}$-bundle P over a Riemann surface Σ, and let $\xi_0 = P \times_{G_{\mathbb{R}}} G$ be the associated principal G-bundle. Then the space $\mathcal{A}(P)$ of $G_{\mathbb{R}}$-connections on P is isomorphic to the space $\mathcal{C}(\xi_0)$ of $(0,1)$-connections ($\bar{\partial}$-operators) on ξ_0 as infinite dimensional complex affine spaces. More explicitly, $\mathcal{A}(P)$ and $\mathcal{C}(\xi_0)$ are affine spaces whose associated vector spaces are $\Omega^1(\Sigma, \mathfrak{g}_{\mathbb{R}})$ and $\Omega^{0,1}(\Sigma, \mathfrak{g})$, respectively, where $\mathfrak{g}_{\mathbb{R}}$ and $\mathfrak{g} = \mathfrak{g}_{\mathbb{R}} \otimes_{\mathbb{R}} \mathbb{C}$ are the Lie algebras of $G_{\mathbb{R}}$ and G, respectively. Choose a local orthonormal frame (θ^1, θ^2) of cotangent bundle T^*_{Σ} of Σ such that $*\theta^1 = \theta^2$. Define an isomorphism $j : \Omega^1(\Sigma, \mathfrak{g}_{\mathbb{R}}) \to \Omega^{0,1}(\Sigma, \mathfrak{g})$ by

$$j(X_1 \otimes \theta^1 + X_2 \otimes \theta^2) = (X_1 + \sqrt{-1}X_2) \otimes (\theta^1 - \sqrt{-1}\theta^2)$$

where $X_1, X_2 \in \Omega^0(\Sigma, \mathfrak{g}_{\mathbb{R}})$. It is easily checked that the definition is independent of choice of (θ^1, θ^2).

Harder and Narasimhan [**HN**] defined a stratification on $\mathcal{C}(\xi_0)$ when $G = GL(n, \mathbb{C})$, and Ramanathan [**Ra**] extended this to general reductive groups. It was conjectured by Atiyah and Bott in [**AB**], and proved by Daskalopoulos in [**Da**] (see also [**Rå**]), that under the isomorphism $\mathcal{A}(P) \cong \mathcal{C}(\xi_0)$, the stratification on $\mathcal{C}(\xi_0)$ coincides with the Morse stratification of the Yang-Mills functional on $\mathcal{A}(P)$.

In this chapter, we first review the description of the stratification in terms of Atiyah-Bott points, following [**AB**, Section 10] and [**FM**]. Then we write down the Atiyah-Bott points for classical groups explicitly, similar to the description of the stratification in terms of slopes when $G_{\mathbb{R}} = U(n)$.

3.1. Preliminaries on reductive Lie groups and Lie algebras

We have

$$\mathfrak{g} = \mathfrak{z}_G \oplus [\mathfrak{g}, \mathfrak{g}]$$

where $\mathfrak{z}_G$ is the center of $\mathfrak{g}$ and $[\mathfrak{g}, \mathfrak{g}]$ is the maximal semisimple subalgebra of $\mathfrak{g}$. Let $H_{\mathbb{R}}$ be a maximal torus of $G_{\mathbb{R}}$, and let $\mathfrak{h}_{\mathbb{R}}$ be the Lie algebra of $H_{\mathbb{R}}$. Then $\mathfrak{h} = \mathfrak{h}_{\mathbb{R}} \otimes_{\mathbb{R}} \mathbb{C}$ is a Cartan subalgebra of $\mathfrak{g}$. Recall that any two maximal tori of $G_{\mathbb{R}}$ are conjugate to each other, and any two Cartan subalgebras of $\mathfrak{g}$ are conjugate to each other. We have $\mathfrak{h} = \mathfrak{z}_G \oplus \mathfrak{h}'$ where $\mathfrak{h}' = \mathfrak{h} \cap [\mathfrak{g}, \mathfrak{g}]$. Here we fix a choice of $H_{\mathbb{R}}$, or equivalently, we fix a Cartan subalgebra $\mathfrak{h}$ of $\mathfrak{g}$. Let R be the root system associated to $\mathfrak{h}$. We have

$$\mathfrak{g} = \mathfrak{h} \oplus \bigoplus_{\alpha \in R} \mathfrak{g}_\alpha = \mathfrak{z}_G \oplus \mathfrak{h}' \oplus \bigoplus_{\alpha \in R} \mathfrak{g}_\alpha.$$

We choose a system of simple roots $\Delta \subset R$, and let R_+ be the set of positive roots. The *Borel subalgebra* associated to Δ is given by

$$\mathfrak{b} = \mathfrak{h} \oplus \bigoplus_{\alpha \in R_+} \mathfrak{g}_\alpha.$$

The Lie algebra of a Borel subgroup B of G is a Borel subalgebra of $\mathfrak{g}$. We have $B \cap G_\mathbb{R} = H_\mathbb{R}$.

A *parabolic subgroup* P of G is a subgroup containing a Borel subgroup, and a *parabolic subalgebra* $\mathfrak{p}$ of $\mathfrak{g}$ is a subalgebra containing a Borel subalgebra. A parabolic subalgebra containing $\mathfrak{b}$ is of the form

$$\mathfrak{p} = \mathfrak{h} \oplus \bigoplus_{\alpha \in \Gamma} \mathfrak{g}_\alpha$$

where

$$\Gamma = R_+ \cup \{\alpha \in R \mid \alpha \in \operatorname{span}(\Delta - I)\}. \tag{3.1}$$

for some subset I of the set Δ of simple roots. There is a one-to-one correspondence between any two of the following:

(i) Subsets $I \subseteq \Delta$.
(ii) Parabolic subalgebras containing a fixed Borel subalgebra $\mathfrak{b}$.
(iii) Parabolic subgroups containing a fixed Borel subgroup B.

In particular, I being the empty set corresponds to G (or $\mathfrak{g}$), and I being the entire set Δ corresponds to B (or $\mathfrak{b}$).

Given a parabolic subalgebra

$$\mathfrak{p} = \mathfrak{h} \oplus \bigoplus_{\alpha \in \Gamma} \mathfrak{g}_\alpha,$$

with Γ as in (3.1), define $-\Gamma$ to be the set of negatives of the members of Γ. In other words, $-\Gamma = -R_+ \cup \{\alpha \in R \mid \alpha \in \operatorname{span}(\Delta - I)\}$. let

$$\mathfrak{l} = \mathfrak{h} \oplus \bigoplus_{\alpha \in \Gamma \cap -\Gamma} \mathfrak{g}_\alpha, \quad \mathfrak{u} = \bigoplus_{\alpha \in \Gamma, \alpha \notin -\Gamma} \mathfrak{g}_\alpha$$

so that $\mathfrak{p} = \mathfrak{l} \oplus \mathfrak{u}$. Then $\mathfrak{l}$, $\mathfrak{u}$ are subalgebras of $\mathfrak{p}$ and $\mathfrak{u}$ is an ideal of $\mathfrak{p}$. The subalgebra $\mathfrak{u}$ is nilpotent, and is called the *nilpotent radical* of $\mathfrak{p}$. The subalgebra $\mathfrak{l}$ is reductive, and is called the *Levi factor* of $\mathfrak{p}$. Let P be the parabolic subgroup with Lie algebra $\mathfrak{p}$. Let $P = LU$ be the semi-direct product associated to the direct sum $\mathfrak{p} = \mathfrak{l} \oplus \mathfrak{u}$, so that the Lie algebras of L and U are $\mathfrak{l}$ and $\mathfrak{u}$, respectively. The reductive Lie group L is called the *Levi factor* of P, and U is called the *unipotent radical* of P. We have $P \cap G_\mathbb{R} = L_\mathbb{R}$, the maximal compact subgroup of L; L is the complexification of $L_\mathbb{R}$.

For simple Lie groups, there is a one-to-one correspondence between simple roots and nodes of the Dynkin diagram. In particular, a (proper) maximal parabolic subgroup corresponds to omitting one node of the Dynkin diagram. See for example [**FH**, Lecture 23].

(A) $G_\mathbb{R} = SU(n)$, $G = SL(n, \mathbb{C})$, $n \geq 2$.
The Dynkin diagram of $\mathfrak{sl}(n, \mathbb{C})$ is A_{n-1}. Omitting a node of A_{n-1}, we get the disjoint union of A_{n_1-1} and A_{n_2-1}, where $n_1 + n_2 = n$, $n_1, n_2 \geq 1$ (with the convention that A_0 is empty). The corresponding parabolic subgroup

P of $SL(n, \mathbb{C})$ is the subgroup which leaves the subspace $\mathbb{C}^{n_1} \times \{0\}$ of $\mathbb{C}^n$ invariant. We have

$$P \cap SU(n) = \{\mathrm{diag}(A, B) \mid A \in U(n_1), B \in U(n_2), \det(A)\det(B) = 1\}.$$

For a general parabolic subgroup P of $SL(n, \mathbb{C})$, we have

$$P \cap SU(n) = \{A \in U(n_1) \times \cdots \times U(n_r) \mid \det(A) = 1\}$$

corresponding to omitting $(r-1)$ nodes, where $n_1 + \cdots + n_r = n$, $n_i \geq 1$.

(B) $G_{\mathbb{R}} = SO(2n+1)$, $G = SO(2n+1, \mathbb{C})$, $n \geq 1$.
The Dynkin Diagram of $\mathfrak{so}(2n+1, \mathbb{C})$ is B_n (with the convention $B_1 = A_1$). Omitting a node of B_n, we get the disjoint union of A_{n_1-1} and B_{n_2}, where $n_1 + n_2 = n$, $n_1 \geq 1$, $n_2 \geq 0$ (with the convention that B_0 is empty). The corresponding parabolic subgroup of $SO(2n+1, \mathbb{C})$ is the subgroup which leaves the following n_1-dimensional subspace of $\mathbb{C}^{2n+1}$ invariant:

$$\{(z_1, \sqrt{-1}z_1, \ldots, z_{n_1}, \sqrt{-1}z_{n_1}, 0, \ldots, 0) \mid z_1, \ldots, z_{n_1} \in \mathbb{C}\}.$$

We have

$$P \cap SO(2n+1) \cong U(n_1) \times SO(2n_2+1).$$

For a general parabolic subgroup P of $SO(2n+1, \mathbb{C})$, we have

$$P \cap SO(2n+1) \cong U(n_1) \times \cdots \times U(n_{r-1}) \times SO(2n_r+1)$$

corresponding to omitting $(r-1)$ nodes, where $n_1 + \cdots + n_r = n$, $n_i \geq 1$ for $i \neq r$, and $n_r \geq 0$ (with the convention that $SO(1)$ is the trivial group).

(C) $G_{\mathbb{R}} = Sp(n)$, $G = Sp(n, \mathbb{C})$, $n \geq 1$.
The Dynkin diagram of $\mathfrak{sp}(n, \mathbb{C})$ is C_n (with the convention $C_1 = A_1$). Omitting a node from C_n, we get the disjoint union of A_{n_1-1} and C_{n_2}, where $n_1 + n_2 = n$, $n_1 \geq 1$, $n_2 \geq 0$ (with the convention that C_0 is the empty set). The corresponding parabolic subgroup of $Sp(n, \mathbb{C})$ is the subgroup which leaves the subspace $\mathbb{C}^{n_1} \times \{0\}$ of $\mathbb{C}^{2n}$ invariant. We have

$$P \cap Sp(n) \cong U(n_1) \times Sp(n_2).$$

For a general parabolic subgroup P of $Sp(n, \mathbb{C})$, we have

$$P \cap Sp(n) \cong U(n_1) \times \cdots \times U(n_{r-1}) \times Sp(n_r)$$

corresponding to omitting $(r-1)$ nodes, where $n_1 + \cdots + n_r = n$, $n_i \geq 1$ for $i \neq r$, and $n_r \geq 0$ (with the convention that $Sp(0)$ is the trivial group).

(D) $G_{\mathbb{R}} = SO(2n)$, $G = SO(2n, \mathbb{C})$, $n \geq 1$.
The Dynkin diagram of $\mathfrak{so}(2n, \mathbb{C})$ is D_n (with the convention $D_1 = A_1$, $D_2 = A_1 \times A_1$, $D_3 = A_3$). Omitting a node of D_n, we get the disjoint union of A_{n_1-1} and D_{n_2}, where $n_1 + n_2 = n$, $n_1 \geq 1$, $n_2 \geq 0$ (with the convention that D_0 is empty). The corresponding parabolic subgroup of $SO(2n, \mathbb{C})$ is the subgroup which leaves the following n_1-dimensional subspace of $\mathbb{C}^{2n}$ invariant:

$$\{(z_1, \sqrt{-1}z_1, \ldots, z_{n_1}, \sqrt{-1}z_{n_1}, 0, \ldots, 0) \mid z_1, \ldots, z_{n_1} \in \mathbb{C}\}.$$

We have

$$P \cap SO(2n) \cong U(n_1) \times SO(2n_2).$$

For a general parabolic subgroup P of $SO(2n, \mathbb{C})$, we have

$$P \cap SO(2n) \cong U(n_1) \times \cdots \times U(n_{r-1}) \times SO(2n_r)$$

corresponding to omitting $(r-1)$ nodes, where $n_1+\cdots+n_r = n$, $n_i \geq 1$ for $i \neq r$, and $n_r \geq 0$ (with the convention that $SO(0)$ is the trivial group). Note that $SO(2) = U(1)$.

3.2. Harder-Narasimhan filtrations of dual vector bundles

Let E be a holomorphic vector bundle over Σ, and let

$$0 = E_0 \subset E_1 \subset \cdots \subset E_r = E$$

be the Harder-Narasimhan filtration, where $D_j = E_j/E_{j-1}$ is semi-stable, and the slopes $\mu_j = \deg(D_j)/\mathrm{rank}(D_j)$ satisfy $\mu_1 > \cdots > \mu_r$. The vector $\mu = (\mu_1, \ldots, \mu_r)$ is the type of E. Let $\mathbb{I}$ denote the trivial holomorphic line bundle over Σ, and let $E^\vee = \mathrm{Hom}(E, \mathbb{I})$ be the dual vector bundle, so that

$$E^\vee_x = \mathrm{Hom}(E_x, \mathbb{C}).$$

Define the subbundle $E^\vee_{-j}$ of $E^\vee$ by

$$(E^\vee_{-j})_x = \{\alpha \in E^\vee_x \mid \alpha(v) = 0 \ \forall v \in (E_j)_x\}.$$

then $(E^\vee_{-j})_x = (E_x/(E_j)_x)^\vee$, and we have

$$0 = E^\vee_{-r} \subset E^\vee_{-r+1} \subset \cdots \subset E^\vee_{-1} \subset E^\vee_0 = E^\vee$$

Let $F_j = E^\vee_{-r+j}/E^\vee_{-r+j-1}$. Then $\mathrm{rank} F_j = \mathrm{rank} D_{r+1-j}$, $\deg F_j = -\deg D_{r+1-j}$, so $\mu(F_j) = -\mu(D_{r+1-j}) = -\mu_{r+1-j}$. The type of $E^\vee$ is given by $(-\mu_r, \ldots, -\mu_1)$, where $-\mu_r > \cdots > -\mu_1$.

3.3. Atiyah-Bott points

Let ξ be a holomorphic principal G-bundle over a Riemann surface, and let $E = \mathrm{ad}\xi = \xi \times_G \mathfrak{g}$ be the associated adjoint bundle.

The Lie algebra $\mathfrak{g}$ has a nondegenerate invariant quadratic form $\mathfrak{g} \to \mathbb{C}$. Therefore, there is a nondegenerate invariant quadratic form I on E, which implies E is self-dual $E^\vee = E$. So the Harder-Narasimhan filtration of E is of the form

$$0 \subset E_{-r} \subset E_{-r+1} \subset \cdots \subset E_{-1} \subset E_0 \subset E_1 \subset \cdots \subset E_{r-1} \subset E.$$

where

$$(E_{-j})_x = \{v \in E_x \mid I(u, v) = 0 \ \forall u \in (E_{j-1})_x\}$$

and $D_0 = E_0/E_{-1}$ has slope zero. Then E_0 is a parabolic subbundle of the Lie algebra bundle E. The structure group G of ξ can then be reduced to a parabolic subgroup Q, such that $\xi = \xi_Q \times_Q G$, where ξ_Q is a holomorphic principal Q-bundle with $\mathrm{ad}\xi_Q = E_0$. The parabolic group is unique up to conjugation, and there is a canonical choice for a fixed Borel subgroup B. This choice gives the *Harder-Narasimhan reduction* and Q is called the *Harder-Narasimhan parabolic* of ξ.

The stratification of the space of holomorphic structures on a fixed topological principal G-bundle ξ is determined by the Harder-Narasimhan parabolic Q together with the topological type of the underlying principal Q-bundle which is an element in $\pi_1(Q)$. To make this more explicit, we describe the stratification in terms of *Atiyah-Bott points*, following [**FM**, Section 2].

Let H be a Cartan subgroup of G. Then $\pi_1(H)$ can be viewed as a lattice in $\sqrt{-1}\mathfrak{h}_{\mathbb{R}}$ such that $\pi_1(H) \otimes_{\mathbb{Z}} \mathbb{R} = \sqrt{-1}\mathfrak{h}_{\mathbb{R}}$.

$$\pi_1(H) \cong \{X \in \sqrt{-1}\mathfrak{h}_{\mathbb{R}} \mid \exp(2\pi\sqrt{-1}X) = e\} \subset \sqrt{-1}\mathfrak{h}_{\mathbb{R}}.$$

For example, $G_{\mathbb{R}} = U(n)$, $\mathfrak{h}_{\mathbb{R}} = \{2\pi\sqrt{-1}\mathrm{diag}(t_1, \ldots, t_n) \mid t_1, \ldots, t_n \in \mathbb{R}\}$, and $\pi_1(H)$ can be identified with the lattice $\{\mathrm{diag}(k_1, \ldots, k_n) \mid k_1, \ldots, k_n \in \mathbb{Z}\} \subset \sqrt{-1}\mathfrak{h}_{\mathbb{R}}$.

The set $\Delta^\vee$ of simple coroots span a sublattice Λ of $\pi_1(H)$, and $\pi_1(G) = \pi_1(H)/\Lambda$. The lattice Λ is called the *coroot lattice* of G. Let $\widehat{\Lambda}$ be the saturation of Λ in $\pi_1(H)$. Then $\pi_1(G_{ss}) \cong \widehat{\Lambda}/\Lambda$. Under the above identification, the short exact sequence of abelian groups

$$1 \to \pi_1(G_{ss}) \to \pi_1(G) \to \pi_1(G/G_{ss}) \to 1$$

can be rewritten as

$$0 \to \widehat{\Lambda}/\Lambda \to \pi_1(H)/\Lambda \to \pi_1(H)/\widehat{\Lambda} \to 0,$$

where $\widehat{\Lambda}/\Lambda$ is a finite abelian group, and $\pi_1(H)/\widehat{\Lambda}$ is a lattice. Let Z_0 denote the connected component of the center of G containing identity. Then $D = Z_0 \cap G_{ss}$ is a finite abelian group, and $G/G_{ss} \cong Z_0/D$. $\pi_1(G/G_{ss}) = \pi_1(H)/\widehat{\Lambda}$ can be identified with a lattice in $\sqrt{-1}\mathfrak{z}_{G_{\mathbb{R}}}$, where $\mathfrak{z}_{G_{\mathbb{R}}} = \mathfrak{z}_G \cap \mathfrak{h}_{\mathbb{R}}$, such that $\pi_1(G/G_{ss}) \otimes_{\mathbb{Z}} \mathbb{R} = \sqrt{-1}\mathfrak{z}_{G_{\mathbb{R}}}$.

Let ξ_0 be a principal G-bundle over a Riemann surface Σ. Its topological type is classified by the second obstruction class $c_1(\xi_0) \in H^2(\Sigma; \pi_1(G)) \cong \pi_1(G)$. Let

$$\mu(\xi_0) \in \pi_1(G/G_{ss}) \subset \sqrt{-1}\mathfrak{z}_{G_{\mathbb{R}}}$$

be the image of $c_1(\xi_0)$ under the projection

$$\pi_1(G) = \pi_1(H)/\Lambda \to \pi_1(G/G_{ss}) = \pi_1(H)/\widehat{\Lambda}.$$

The group $\hat{G} = \mathrm{Hom}(G, \mathbb{C}^*) = \mathrm{Hom}(G/G_{ss}, \mathbb{C}^*)$ can be identified with the dual lattice of $\pi_1(H)/\widehat{\Lambda}$.

Let P^I be a parabolic subgroup determined by $I \subseteq \Delta$, and let L^I be its Levi factor. The topological type of a principal L^I bundle η_0 is determined by $c_1(\eta_0) \in \pi_1(L)$. Given $\xi_0 \in \mathrm{Prin}_G(\Sigma)$, we want to enumerate

$$\{\eta_0 \in \mathrm{Prin}_{L^I}(\Sigma) \mid \eta_0 \times_{L^I} G = \xi_0\}. \tag{3.2}$$

Consider the commutative diagram

$$\begin{array}{ccccccc}
 & & 0 & & 0 & & \\
 & & \downarrow & & \downarrow & & \\
0 & \longrightarrow & \pi_1(L_{ss}) = \hat{\Lambda}_L/\Lambda_L & \xrightarrow{j_{ss}} & \pi_1(G_{ss}) = \widehat{\Lambda}/\Lambda & \xrightarrow{\oplus_{\alpha\in I}\varpi_\alpha} & \oplus_{\alpha\in I}\mathbb{Q}/\mathbb{Z} \\
 & & {\scriptstyle i_L}\downarrow & & {\scriptstyle i_G}\downarrow & & \parallel \\
 & & \pi_1(L) = \pi_1(H)/\Lambda_L & \xrightarrow{j} & \pi_1(G) = \pi_1(H)/\Lambda & \xrightarrow{\oplus_{\alpha\in I}\varpi_\alpha} & \oplus_{\alpha\in I}\mathbb{Q}/\mathbb{Z} \\
 & & {\scriptstyle p_L}\downarrow & & {\scriptstyle p_G}\downarrow & & \\
 & & \pi_1(L/L_{ss}) = \pi_1(H)/\widehat{\Lambda}_L & \xrightarrow{p} & \pi_1(G/G_{ss}) = \pi_1(H)/\widehat{\Lambda} & & \\
 & & \downarrow & & \downarrow & & \\
 & & 0 & & 0 & &
\end{array}$$

where ϖ_α are the fundamental weights. In the above diagram, the columns and the first row are exact.

Given a principal L-bundle η_0, $c_1(\eta_0)\in\pi_1(L)$ is determined by

$$j(c_1(\eta_0)) = c_1(\eta_0\times_L G)\in\pi_1(G),\quad p_L(c_1(\eta_0)) = \mu(\eta_0)\in\pi_1(L/L_{ss}).$$

Given $\xi_0\in \mathrm{Prin}_G(\Sigma)$, we have $c_1(\xi_0)\in\pi_1(G)$ and $\mu(\xi_0)\in\pi_1(G/G_{ss})$. The map p_L restricts to a bijection $j^{-1}(c_1(\xi_0))\to p^{-1}(\mu(\xi_0))$. Note that the set in (3.2) can be identified with $j^{-1}(c_1(\xi_0))$.

LEMMA 3.1 ([**FM**, Lemma 2.1.2]). *Suppose that η_0 is a reduction of ξ_0 to a standard parabolic group P^I for some $I\subseteq\Delta$, possibly empty. The Atiyah-Bott point $\mu(\eta_0)$ and the topological type of ξ_0 as a G-bundle determine the topological type of η_0/U^I as an L^I-bundle (and hence of η_0 as a P^I bundle). Given a point $\mu\in\mathfrak{h}_{\mathbb{R}}$, there is a reduction of ξ to a P^I-bundle whose Atiyah-Bott point is μ if and only if the following conditions hold:*

(i) $\mu\in\sqrt{-1}\mathfrak{z}_{L^I{}_{\mathbb{R}}}$, *where* $\mathfrak{z}_{L^I{}_{\mathbb{R}}}$ *is the center of the Lie algebra of* $L^I_{\mathbb{R}} = L^I\cap G_{\mathbb{R}}$.
(ii) *For every simple root* $\alpha\in I$ *we have* $\varpi_\alpha(\mu)\equiv\varpi_\alpha(c)$ *(mod* $\mathbb{Z}$*).*
(iii) $\chi(\mu)=\chi(c)$ *for all characters* χ *of* G.

DEFINITION 3.2 ([**FM**, Definition 2.1.3]). A pair (μ, I) consisting of a point $\mu\in\sqrt{-1}\mathfrak{h}_{\mathbb{R}}$ and a subset $I\subseteq\Delta$ is said to be of *Atiyah-Bott type* for $c\in\pi_1(G)$ (or ξ_0 where $c_1(\xi_0)=c$) if (i)-(iii) hold. A point $\mu\in\sqrt{-1}\mathfrak{h}_{\mathbb{R}}$ is said to be of *Atiyah-Bott type* for c if there is $I\subseteq\Delta$ such that (μ,I) is a pair of Atiyah-Bott type for c.

One may assume $\mu\in\overline{C}_0$, where $\overline{C}_0$ is the closure of the fundamental Weyl chamber

$$C_0=\{X\in\sqrt{-1}\mathfrak{h}_{\mathbb{R}}\mid \alpha(X)>0\ \forall\alpha\in\Delta\}.$$

We may choose the minimal I such that $\alpha(\mu)>0$ for all $\alpha\in I$. Then the stratum $\mathcal{C}_\mu$ of the space of $(0,1)$-connections on ξ_0 are indexed by points μ of Atiyah-Bott type of $c_1(\xi_0)$ such that $\mu\in\overline{C}_0$. We may incorporate this by adding

(iv) $\alpha(\mu)>0$ for all $\alpha\in I$.

Let $\mathcal{C}(\xi_0)$ be the space of all $(0,1)$-connections defining holomorphic structures on a principal G-bundle ξ_0 with $c_1(\xi_0)=c\in\pi_1(G)$. As a summary of the above discussion, we have following description of the Harder-Narasimhan stratification of $\mathcal{C}$.

DEFINITION 3.3. Given a point $\mu\in\overline{C}_0$ of Atiyah-Bott type for c, the stratum $\mathcal{C}_\mu\subset\mathcal{C}(\xi_0)$ is the set of all $(0,1)$-connections defining holomorphic structures on ξ_0 whose Harder-Narasimhan reduction has Atiyah-Bott type equal to μ. The strata are preserved by the action of gauge group. The union of these strata over all $\mu\in\overline{C}_0$ of Atiyah-Bott type for ξ_0 is $\mathcal{C}(\xi_0)$.

3.4. Atiyah-Bott points for classical groups

In this section, we assume

$$n_1,\dots,n_r\in\mathbb{Z}_{>0},\quad n_1+\cdots+n_r=n.$$

3.4.1. $G_{\mathbb{R}}=U(n)$. $G=GL(n,\mathbb{C})$, and

$$\sqrt{-1}\mathfrak{h}_{\mathbb{R}}=\{\mathrm{diag}(t_1,\dots,t_n)\mid t_i\in\mathbb{R}\}.$$

Let $e_i \in \sqrt{-1}\mathfrak{h}_{\mathbb{R}}$ be defined by $t_j = \delta_{ij}$. Then $\{e_1, \ldots, e_n\}$ is a basis of $\sqrt{-1}\mathfrak{h}_{\mathbb{R}}$. Let $\{\theta_1, \ldots, \theta_n\}$ be the dual basis of $(\sqrt{-1}\mathfrak{h}_{\mathbb{R}})^\vee = \mathrm{Hom}_{\mathbb{R}}(\sqrt{-1}\mathfrak{h}_{\mathbb{R}}, \mathbb{R})$. Then

$$\begin{aligned}
&\pi_1(H) = \mathbb{Z}e_1 \oplus \cdots \oplus \mathbb{Z}e_n \subset \sqrt{-1}\mathfrak{h}_{\mathbb{R}} \\
&\Delta = \{\alpha_i = \theta_i - \theta_{i+1} \mid i = 1, \ldots, n-1\} \subset (\sqrt{-1}\mathfrak{h}_{\mathbb{R}})^\vee \\
&\Delta^\vee = \{\alpha_i^\vee = e_i - e_{i+1} \mid i = 1, \ldots, n-1\} \subset \sqrt{-1}\mathfrak{h}_{\mathbb{R}}
\end{aligned}$$

$\pi_1(U(n)) \cong \pi_1(GL(n, \mathbb{C})) \cong \pi_1(H)/\Lambda \cong \mathbb{Z}$ is generated by e_1 (mod Λ). Let $c = ke_1$ (mod Λ). Then μ satisfies (i)-(iv) in Section 3.3 if and only if

$$\mu = \mathrm{diag}\Big(\frac{k_1}{n_1} I_{n_1}, \ldots, \frac{k_r}{n_r} I_{n_r}\Big)$$

where

$$k_1, \ldots, k_r \in \mathbb{Z}, \quad k_1 + \cdots + k_r = k, \quad \frac{k_1}{n_1} > \frac{k_1}{n_2} > \cdots > \frac{k_r}{n_r}.$$

3.4.2. $G_{\mathbb{R}} = SO(2n+1)$. $G = SO(2n+1, \mathbb{C})$, and

$$\sqrt{-1}\mathfrak{h}_{\mathbb{R}} = \{\sqrt{-1}\mathrm{diag}(t_1 J, \ldots, t_n J, 0I_1) \mid t_i \in \mathbb{R}\}$$

where

$$J = \begin{pmatrix} 0 & -1 \\ 1 & 0 \end{pmatrix}$$

Let $e_i \in \sqrt{-1}\mathfrak{h}_{\mathbb{R}}$ be defined by $t_j = \delta_{ij}$. Then $\{e_1, \ldots, e_n\}$ is a basis of $\sqrt{-1}\mathfrak{h}_{\mathbb{R}}$. Let $\{\theta_1, \ldots, \theta_n\}$ be the dual basis of $(\sqrt{-1}\mathfrak{h}_{\mathbb{R}})^\vee$. Then

$$\begin{aligned}
&\pi_1(H) = \mathbb{Z}e_1 \oplus \cdots \oplus \mathbb{Z}e_n \subset \sqrt{-1}\mathfrak{h}_{\mathbb{R}} \\
&\Delta = \{\alpha_i = \theta_i - \theta_{i+1} \mid i = 1, \ldots, n-1\} \cup \{\alpha_n = \theta_n\} \subset (\sqrt{-1}\mathfrak{h}_{\mathbb{R}})^\vee \\
&\Delta^\vee = \{\alpha_i^\vee = e_i - e_{i+1} \mid i = 1, \ldots, n-1\} \cup \{\alpha_n^\vee = 2e_n\} \subset \sqrt{-1}\mathfrak{h}_{\mathbb{R}}
\end{aligned}$$

$\pi_1(SO(2n+1)) \cong \pi_1(SO(2n+1, \mathbb{C}) \cong \mathbb{Z}/2\mathbb{Z}$ is generated by e_n (mod Λ). $c = ke_n$ (mod Λ) corresponds to $w_2 = k$ where $k = 0, 1$.

Case 1. $\alpha_n \in I$. Then μ satisfies (i)-(iv) in Section 3.3 if and only if

$$\mu = \sqrt{-1}\mathrm{diag}\Big(\frac{k_1}{n_1} J_{n_1}, \ldots, \frac{k_r}{n_r} J_{n_r}, 0I_1\Big)$$

where

$$k_1, \ldots, k_r \in \mathbb{Z}, \quad k_1 + \cdots + k_r \equiv k \ (\mathrm{mod}\ 2\mathbb{Z}), \quad \frac{k_1}{n_1} > \frac{k_2}{n_2} > \cdots \frac{k_r}{n_r} > 0.$$

Case 2. $\alpha_n \notin I$. Then μ satisfies (i)-(iv) in Section 3.3 if and only if

$$\mu = \sqrt{-1}\mathrm{diag}\Big(\frac{k_1}{n_1} J_{n_1}, \ldots, \frac{k_{r-1}}{n_{r-1}} J_{n_{r-1}}, 0I_{2n_r+1}\Big)$$

where

$$k_1, \ldots, k_{r-1} \in \mathbb{Z}, \quad \frac{k_1}{n_1} > \frac{k_2}{n_2} > \cdots > \frac{k_{r-1}}{n_{r-1}} > 0.$$

3.4.3. $G_{\mathbb{R}} = SO(2n)$. $G = SO(2n, \mathbb{C})$, and

$$\sqrt{-1}\mathfrak{h}_{\mathbb{R}} = \{\sqrt{-1}\mathrm{diag}(t_1 J, \dots, t_n J) \mid t_i \in \mathbb{R}\}$$

where

$$J = \begin{pmatrix} 0 & -1 \\ 1 & 0 \end{pmatrix}$$

Let $e_i \in \sqrt{-1}\mathfrak{h}_{\mathbb{R}}$ be defined by $t_j = \delta_{ij}$. Then $\{e_1, \dots, e_n\}$ is a basis of $\sqrt{-1}\mathfrak{h}_{\mathbb{R}}$. Let $\{\theta_1, \dots, \theta_n\}$ be the dual basis of $(\sqrt{-1}\mathfrak{h}_{\mathbb{R}})^\vee$. Then

$$\pi_1(H) = \mathbb{Z}e_1 \oplus \cdots \oplus \mathbb{Z}e_n \subset \sqrt{-1}\mathfrak{h}_{\mathbb{R}}$$
$$\Delta = \{\alpha_i = \theta_i - \theta_{i+1} \mid i = 1, \dots, n-1\} \cup \{\alpha_n = \theta_{n-1} + \theta_n\} \subset (\sqrt{-1}\mathfrak{h}_{\mathbb{R}})^\vee$$
$$\Delta^\vee = \{\alpha_i^\vee = e_i - e_{i+1} \mid i = 1, \dots, n-1\} \cup \{\alpha_n^\vee = e_{n-1} + e_n\} \subset \sqrt{-1}\mathfrak{h}_{\mathbb{R}}$$

$\pi_1(SO(2n)) \cong \pi_1(SO(2n, \mathbb{C})) \cong \mathbb{Z}/2\mathbb{Z}$ is generated by e_n (mod Λ). $c = ke_n$ (mod Λ) corresponds to $w_2 = k$ where $k = 0, 1$.

Case 1. $\alpha_{n-1}, \alpha_n \in I$, $n_r = 1$. Then μ satisfies (i)-(iv) in Section 3.3 if and only if

$$\mu = \sqrt{-1}\mathrm{diag}\Big(\frac{k_1}{n_1} J_{n_1}, \dots, \frac{k_{r-1}}{n_{r-1}} J_{n_{r-1}}, k_r J\Big)$$

where

$$k_1, \dots, k_r \in \mathbb{Z}, \quad k_1 + \cdots + k_r \equiv k \ (\mathrm{mod}\ 2\mathbb{Z}), \quad \frac{k_1}{n_1} > \frac{k_2}{n_2} > \cdots > \frac{k_{r-1}}{n_{r-1}} > |k_r|.$$

Case 2. $\alpha_{n-1} \in I$, $\alpha_n \notin I$, $n_r > 1$. Then μ satisfies (i)-(iv) in Section 3.3 if and only if

$$\mu = \sqrt{-1}\mathrm{diag}\Big(\frac{k_1}{n_1} J_{n_1}, \dots, \frac{k_{r-1}}{n_{r-1}} J_{n_{r-1}}, \frac{k_r}{n_r} J_{n_r - 1}, -\frac{k_r}{n_r} J\Big)$$

where

$$k_1, \dots, k_r \in \mathbb{Z}, \quad k_1 + \cdots + k_r = k \ (\mathrm{mod}\ 2\mathbb{Z}), \quad \frac{k_1}{n_1} > \frac{k_2}{n_2} > \cdots > \frac{k_r}{n_r} > 0.$$

Case 3. $\alpha_{n-1} \notin I$, $\alpha_n \in I$, $n_r > 1$. Then μ satisfies (i)-(iv) in Section 3.3 if and only if

$$\mu = \sqrt{-1}\mathrm{diag}\Big(\frac{k_1}{n_1} J_{n_1}, \dots, \frac{k_r}{n_r} J_{n_r}\Big)$$

where

$$k_1, \dots, k_r \in \mathbb{Z}, \quad k_1 + \cdots + k_r = k \ (\mathrm{mod}\ 2\mathbb{Z}), \quad \frac{k_1}{n_1} > \frac{k_2}{n_2} > \cdots > \frac{k_r}{n_r} > 0.$$

Case 4. $\alpha_{n-1} \notin I$, $\alpha_n \notin I$. Then μ satisfies (i)-(iv) in Section 3.3 if and only if

$$\mu = \sqrt{-1}\mathrm{diag}\Big(\frac{k_1}{n_1} J_{n_1}, \dots, \frac{k_{r-1}}{n_{r-1}} J_{n_r}, 0 J_{n_r}\Big)$$

where

$$k_1, \dots, k_{r-1} \in \mathbb{Z}, \quad \frac{k_1}{n_1} > \frac{k_2}{n_2} > \cdots > \frac{k_{r-1}}{n_{r-1}} > 0.$$

3.4.4. $G_{\mathbb{R}} = Sp(n)$. $G = Sp(n, \mathbb{C})$, and

$$\sqrt{-1}\mathfrak{h}_{\mathbb{R}} = \{\mathrm{diag}(t_1, \ldots, t_n, -t_1, \ldots, -t_n) \mid t_i \in \mathbb{R}\}$$

Let $e_i \in \sqrt{-1}\mathfrak{h}_{\mathbb{R}}$ be defined by $t_j = \delta_{ij}$. Then $\{e_1, \ldots, e_n\}$ is a basis of $\sqrt{-1}\mathfrak{h}_{\mathbb{R}}$. Let $\{\theta_1, \ldots, \theta_n\}$ be the dual basis of $(\sqrt{-1}\mathfrak{h}_{\mathbb{R}})^{\vee}$. Then

$$\pi_1(H) = \mathbb{Z}e_1 \oplus \cdots \oplus \mathbb{Z}e_n \subset \sqrt{-1}\mathfrak{h}_{\mathbb{R}}$$
$$\Delta = \{\alpha_i = \theta_i - \theta_{i+1} \mid i = 1, \ldots, n-1\} \cup \{2\theta_n\} \subset (\sqrt{-1}\mathfrak{h}_{\mathbb{R}})^{\vee}$$
$$\Delta^{\vee} = \{\alpha_i^{\vee} = e_i - e_{i+1} \mid i = 1, \ldots, n-1\} \cup \{e_n\} \subset \sqrt{-1}\mathfrak{h}_{\mathbb{R}}$$

$\pi_1(Sp(n)) \cong \pi_1(Sp(n, \mathbb{C}))$ is trivial.

Case 1. $\alpha_n \in I$. Then μ satisfies (i)-(iv) in Section 3.3 if and only of

$$\mu = \mathrm{diag}\Big(\frac{k_1}{n_1}I_{n_1}, \ldots, \frac{k_r}{n_r}I_{n_r}, -\frac{k_1}{n_1}I_{n_1}, \ldots, -\frac{k_r}{n_r}I_{n_r}\Big)$$

where

$$k_1, \ldots, k_r \in \mathbb{Z}, \quad \frac{k_1}{n_1} > \frac{k_2}{n_2} > \cdots > \frac{k_r}{n_r} > 0.$$

Case 2. $\alpha_n \notin I$. Then μ satisfies (i)-(iv) of Section 3.3 if and only if

$$\mu = \mathrm{diag}\Big(\frac{k_1}{n_1}I_{n_1}, \ldots, \frac{k_{r-1}}{n_{r-1}}I_{n_{r-1}}, 0I_{n_r}, -\frac{k_1}{n_1}I_{n_1}, \ldots, -\frac{k_{r-1}}{n_{r-1}}I_{n_{r-1}}, 0I_{n_r}\Big)$$

where

$$k_1, \ldots, k_{r-1} \in \mathbb{Z}, \quad \frac{k_1}{n_1} > \frac{k_2}{n_2} > \cdots > \frac{k_{r-1}}{n_{r-1}} > 0.$$

CHAPTER 4

Yang-Mills Connections and Representation Varieties

Let $G_{\mathbb{R}}$ be a compact connected Lie group, and let P be a C^∞ principal $G_{\mathbb{R}}$-bundle over a closed (orientable or nonorientable) surface. In [**HL4**, Section 3], we introduced Yang-Mills functional and Yang-Mills connections on closed nonorientable surfaces.

In this chapter, we study the connected components of the representation variety of Yang-Mills connections. We recover the description of the Morse stratification in terms of Atiyah-Bott points for orientable Σ (Section 4.2), and determine candidates of Atiyah-Bott points for nonorientable Σ (Section 4.5). We also discuss and give a closed formula for $G_{\mathbb{R}}$-equivariant rational Poincaré series of the representation variety of central Yang-Mills connections (Section 4.3). In Section 4.6 and Section 4.7, we introduce certain twisted representation varieties that will arise in Chapter 5, Chapter 6, and Chapter 7, and study their connectedness.

4.1. Representation varieties for Yang-Mills connections

Let $\mathcal{A}(P)$ be the space of $G_{\mathbb{R}}$-connections on P, and let $\mathcal{N}(P)$ be the space of Yang-Mills connections on P. Let $\mathcal{G}(P) = \mathrm{Aut}(P)$ be the gauge group, and let $\mathcal{G}_0(P)$ be the base gauge group. Let $\Gamma_{\mathbb{R}}(\Sigma)$ be the super central extension of $\pi_1(\Sigma)$ defined in [**HL4**, Section 4.1].

THEOREM 4.1 ([**AB**, Theorem 6.7], [**HL4**, Theorem 4.6]). *There is a bijective correspondence between conjugacy classes of homomorphisms $\Gamma_{\mathbb{R}}(\Sigma) \to G_{\mathbb{R}}$ and gauge equivalence classes of Yang-Mills $G_{\mathbb{R}}$-connections over Σ. In other words,*

$$\bigcup_{P\in \mathrm{Prin}_{G_{\mathbb{R}}}(\Sigma)} \mathcal{N}(P)/\mathcal{G}_0(P) \cong \mathrm{Hom}(\Gamma_{\mathbb{R}}(\Sigma), G_{\mathbb{R}})$$

$$\bigcup_{P\in \mathrm{Prin}_{G_{\mathbb{R}}}(\Sigma)} \mathcal{N}(P)/\mathcal{G}(P) \cong \mathrm{Hom}(\Gamma_{\mathbb{R}}(\Sigma), G_{\mathbb{R}})/G_{\mathbb{R}}$$

To describe $\mathrm{Hom}(\Gamma_{\mathbb{R}}(\Sigma), G_{\mathbb{R}})$ more explicitly, we introduce some notation. Let Σ_0^ℓ be the closed, compact, connected, orientable surface with $\ell \geq 0$ handles. Let Σ_1^ℓ be the connected sum of Σ_0^ℓ and $\mathbb{RP}^2$, and let Σ_2^ℓ be the connected sum of Σ_0^ℓ and a Klein bottle. Any closed, compact, connected surface is of the form Σ_i^ℓ, where ℓ is a nonnegative integer and $i = 0, 1, 2$. Σ_i^ℓ is orientable if and only if $i = 0$. Let $(G_{\mathbb{R}})_X$ denote the stabilizer of X of the adjoint action of $G_{\mathbb{R}}$ on $\mathfrak{g}_{\mathbb{R}}$. With the above notation, $\mathrm{Hom}(\Gamma_{\mathbb{R}}(\Sigma_i^\ell), G_{\mathbb{R}})$ can be identified with the representation variety $X_{\mathrm{YM}}^{\ell,i}(G_{\mathbb{R}})$, where

$$\begin{aligned}
X^{\ell,0}_{\mathrm{YM}}(G_{\mathbb{R}}) &= \{(a_1,b_1,\ldots,a_\ell,b_\ell,X)\in G_{\mathbb{R}}{}^{2\ell}\times\mathfrak{g}_{\mathbb{R}} \mid \\
&\qquad a_i,b_i\in(G_{\mathbb{R}})_X,\ \prod_{i=1}^{\ell}[a_i,b_i]=\exp(X)\}\\
X^{\ell,1}_{\mathrm{YM}}(G_{\mathbb{R}}) &= \{(a_1,b_1,\ldots,a_\ell,b_\ell,c,X)\in G_{\mathbb{R}}{}^{2\ell+1}\times\mathfrak{g}_{\mathbb{R}} \mid \\
&\qquad a_i,b_i\in(G_{\mathbb{R}})_X,\ \mathrm{Ad}(c)X=-X,\ \prod_{i=1}^{\ell}[a_i,b_i]=\exp(X)c^2\}\\
X^{\ell,2}_{\mathrm{YM}}(G_{\mathbb{R}}) &= \{(a_1,b_1,\ldots,a_\ell,b_\ell,d,c,X)\in G_{\mathbb{R}}{}^{2\ell+2}\times\mathfrak{g}_{\mathbb{R}} \mid \\
&\qquad a_i,b_i,d\in(G_{\mathbb{R}})_X,\ \mathrm{Ad}(c)X=-X,\ \prod_{i=1}^{\ell}[a_i,b_i]=\exp(X)cdc^{-1}d\}
\end{aligned}$$

The $G_{\mathbb{R}}$-action on $X^{\ell,i}_{\mathrm{YM}}(G_{\mathbb{R}})$ is given by

$$g\cdot(c_1,\ldots,c_{2\ell+i},X)=(gc_1g^{-1},\ldots,gc_{2\ell+i}g^{-1},\mathrm{Ad}(g)X).$$

4.2. Connected components of the representation variety for orientable surfaces

$G_{\mathbb{R}}$ is connected, so the natural projection

$$X^{\ell,0}_{\mathrm{YM}}(G_{\mathbb{R}})\to X^{\ell,0}_{\mathrm{YM}}(G_{\mathbb{R}})/G_{\mathbb{R}}$$

induces a bijection

$$\pi_0(X^{\ell,0}_{\mathrm{YM}}(G_{\mathbb{R}}))\to\pi_0(X^{\ell,0}_{\mathrm{YM}}(G_{\mathbb{R}})/G_{\mathbb{R}}).$$

Any point in $X^{\ell,0}_{\mathrm{YM}}(G_{\mathbb{R}})/G_{\mathbb{R}}$ can be represented by

$$(a_1,b_1,\ldots,a_\ell,b_\ell,X)$$

where $X\in\mathfrak{h}_{\mathbb{R}}$. Such representative is unique if we require that $\sqrt{-1}X$ is in the closure $\overline{C}_0$ of the fundamental Weyl chamber

$$C_0=\{Y\in\sqrt{-1}\mathfrak{h}_{\mathbb{R}}\mid\alpha(Y)>0,\ \forall\alpha\in R_+\}=\{Y\in\sqrt{-1}\mathfrak{h}_{\mathbb{R}}\mid\alpha(Y)>0,\ \forall\alpha\in\Delta\}.$$

Given X such that $\sqrt{-1}X\in\overline{C}_0$, we want to find the stabilizer $(G_{\mathbb{R}})_X$ of the adjoint action of $G_{\mathbb{R}}$ on $\mathfrak{g}_{\mathbb{R}}$. Let G be the complexification of $G_{\mathbb{R}}$. We use the notation in Chapter 3. Let

$$I_X=\{\alpha\in\Delta\mid\alpha(\sqrt{-1}X)>0\}.$$

Then $I_X=\Delta$ if $\sqrt{-1}X\in C_0$, and I_X is empty if and only if X is in the center $\mathfrak{z}_{G_{\mathbb{R}}}$ of $\mathfrak{g}_{\mathbb{R}}$. Let

$$\Gamma_X=R_+\cup\{\alpha\in R\mid\alpha\in\mathrm{span}(\Delta-I_X)\}.$$

The stabilizer $\mathfrak{g}_X$ of the adjoint action of $\mathfrak{g}$ on itself is the Levi factor of the parabolic subalgebra

$$\mathfrak{p}_X=\mathfrak{h}\oplus\bigoplus_{\alpha\in\Gamma_X}\mathfrak{g}_\alpha.$$

We have $\mathfrak{p}_X=\mathfrak{g}_X\oplus\mathfrak{u}_X$, where $\mathfrak{g}_X$ and $\mathfrak{u}_X$ are the Levi factor and the nilpotent radical of $\mathfrak{p}_X$, respectively. The Lie algebra of G_X is $\mathfrak{g}_X$. We conclude that

$$(G_{\mathbb{R}})_X=L^{I_X}\cap G_{\mathbb{R}}=L^{I_X}_{\mathbb{R}}.$$

Note that

$$X \in \mathfrak{z}_{L_{\mathbb{R}}^{I_X}}, \quad \exp(X) = \prod_{i=1}^{\ell}[a_i, b_i] \in (L_{\mathbb{R}}^{I_X})_{ss}.$$

Let $\mu_X = \frac{\sqrt{-1}}{2\pi}X$. Then

$$\mu_X \in \pi_1(H)/\widehat{\Lambda}_{L^{I_X}} \subset \sqrt{-1}\mathfrak{z}_{L_{\mathbb{R}}^{I_X}} \subset \sqrt{-1}\mathfrak{h}_{\mathbb{R}}$$

and (μ_X, I_X) is of Atiyah-Bott type for some $c \in \pi_1(G) = \pi_1(G_{\mathbb{R}})$.

We now state the condition for $X \in \mathfrak{h}_{\mathbb{R}}$ such that $(a_1, b_1, \ldots, a_\ell, b_\ell, X) \in X_{\mathrm{YM}}^{\ell,0}(G_{\mathbb{R}})$ for some $(a_1, b_1, \ldots, a_\ell, b_\ell) \in G^{2\ell}$. Given $I \subseteq \Delta$, let Z^I be the connected component of the identity of the center of $L_{\mathbb{R}}^I$, and let D^I be the center of $(L_{\mathbb{R}}^I)_{ss}$. Then the Lie algebra for Z^I is $\mathfrak{z}_{L_{\mathbb{R}}^I}$. Denote

$$\begin{aligned} \Xi^I &= \{\mu \in \sqrt{-1}\mathfrak{z}_{L_{\mathbb{R}}^I} \mid \exp(-2\pi\sqrt{-1}\mu) \in D^I\} \cong \pi_1(Z^I/D^I) \cong \pi_1(L_{\mathbb{R}}^I/(L_{\mathbb{R}}^I)_{ss}) \\ \Xi_+^I &= \{\mu \in \Xi^I \cap \overline{C}_0 \mid \alpha(\mu) > 0 \text{ iff } \alpha \in I\}. \end{aligned}$$

Given $\mu \in \Xi_+^I$, let $X_\mu = -2\pi\sqrt{-1}\mu \in \mathfrak{h}_{\mathbb{R}}$. Suppose that $(a_1, b_1, \ldots, a_\ell, b_\ell, X) \in X_{\mathrm{YM}}^{\ell,0}(G_{\mathbb{R}})$. Then there is a unique pair (μ, I), where $I \subseteq \Delta$ and $\mu \in \Xi_+^I$, such that X is conjugate to X_μ. Let $C_\mu \subset \mathfrak{g}_{\mathbb{R}}$ denote the conjugacy class of X_μ, and define

$$X_{\mathrm{YM}}^{\ell,0}(G_{\mathbb{R}})_\mu = \{(a_1, b_1, \ldots, a_\ell, b_\ell, X) \in {G_{\mathbb{R}}}^{2\ell} \times C_\mu \mid a_i,\ b_i \in (G_{\mathbb{R}})_X, \prod_{i=1}^{\ell}[a_i, b_i] = \exp(X)\}.$$

Then $X_{\mathrm{YM}}^{\ell,0}(G_{\mathbb{R}})$ is a disjoint union of

$$\{X_{\mathrm{YM}}^{\ell,0}(G_{\mathbb{R}})_\mu \mid \mu \in \Xi_+^I, I \subseteq \Delta\}.$$

Each $X_{\mathrm{YM}}^{\ell,0}(G_{\mathbb{R}})_\mu$ is a union of finitely many connected components of $X_{\mathrm{YM}}^{\ell,0}(G_{\mathbb{R}})$.

Note that $(G_{\mathbb{R}})_{X_\mu} = L_{\mathbb{R}}^I$ for $\mu \in \Xi_+^I$. We define *reduced representation varieties*

(4.1)

$$V_{\mathrm{YM}}^{\ell,0}(G_{\mathbb{R}})_\mu = \{(a_1, b_1, \ldots, a_\ell, b_\ell) \in (L_{\mathbb{R}}^I)^{2\ell} \mid \prod_{i=1}^{\ell}[a_i, b_i] = \exp(X_\mu)\} \cong X_{\mathrm{YM}}^{\ell,0}(L_{\mathbb{R}}^I)_\mu.$$

They correspond to the reduction from $G_{\mathbb{R}}$ to the subgroup $L_{\mathbb{R}}^I$. More precisely, we have a homeomorphism

$$X_{\mathrm{YM}}^{\ell,0}(G_{\mathbb{R}})_\mu/G_{\mathbb{R}} \cong V_{\mathrm{YM}}^{\ell,0}(G_{\mathbb{R}})_\mu/L_{\mathbb{R}}^I$$

and a homotopy equivalence

$$X_{\mathrm{YM}}^{\ell,0}(G_{\mathbb{R}})_\mu{}^{hG_{\mathbb{R}}} \sim V_{\mathrm{YM}}^{\ell,0}(G_{\mathbb{R}})_\mu{}^{hL_{\mathbb{R}}^I}$$

where X^{hG} denote the homotopic orbit space $EG \times_G X$.

We now recall the formulation in [**HL3**, Section 2.1]. Let $\rho_{ss} : \widetilde{(L_{\mathbb{R}}^I)_{ss}} \to (L_{\mathbb{R}}^I)_{ss}$ be the universal cover. Then the universal cover of $L_{\mathbb{R}}^I$ is given by

$$\rho : \widetilde{L_{\mathbb{R}}^I} = \mathfrak{z}_{L_{\mathbb{R}}^I} \times \widetilde{(L_{\mathbb{R}}^I)_{ss}} \to L_{\mathbb{R}}^I, \quad (X, g) \mapsto \exp_{Z^I}(X)\rho_{ss}(g)$$

where $\exp_{Z^I} : \mathfrak{z}_{L_{\mathbb{R}}^I} \to Z^I$ is the exponential map. We have

$$\pi_1((L_{\mathbb{R}}^I)_{ss}) \cong \mathrm{Ker}(\rho_{ss}), \quad \pi_1(L_{\mathbb{R}}^I) \cong \mathrm{Ker}\rho \subset (-2\pi\sqrt{-1}\Xi^I) \times Z(\widetilde{(L_{\mathbb{R}}^I)_{ss}}) \subset \mathfrak{z}_{L_{\mathbb{R}}^I} \times \widetilde{(L_{\mathbb{R}}^I)_{ss}}.$$

The map

$$p_{L^I_{\mathbb{R}}} : \mathrm{Ker}\rho \to \Xi^I, \quad (X,g) \mapsto \frac{\sqrt{-1}}{2\pi} X$$

coincides with the surjective group homomorphism

$$p_{L^I_{\mathbb{R}}} : \pi_1(L^I_{\mathbb{R}}) \to \pi_1(L^I_{\mathbb{R}}/(L^I_{\mathbb{R}})_{ss})$$

under the isomorphisms $\mathrm{Ker}\rho \cong \pi_1(L^I_{\mathbb{R}})$ and $\Xi^I \cong \pi_1(L^I_{\mathbb{R}}/(L^I_{\mathbb{R}})_{ss})$.

Define the obstruction map $o : V^{\ell,0}_{\mathrm{YM}}(G_{\mathbb{R}})_\mu \to p^{-1}_{L^I_{\mathbb{R}}}(\mu)$ as follows. Given a point $(a_1, b_1, \ldots, a_\ell, b_\ell) \in V^{\ell,0}_{\mathrm{YM}}(G_{\mathbb{R}})_\mu$, choose $\tilde{a}_i \in \rho^{-1}(a_i)$, $\tilde{b}_i \in \rho^{-1}(b_i)$. Define $o(a_1, b_1, \ldots, a_\ell, b_\ell) = \prod_{i=1}^{\ell} [\tilde{a}_i, \tilde{b}_i]$. Note that this definition does not depend on the choice of $\tilde{a}_i, \tilde{b}_i$. We have $o(a_1, b_1, \ldots, a_\ell, b_\ell) \in \{0\} \times \widetilde{(L^I_{\mathbb{R}})_{ss}}$, and

$$\rho_{ss}(o(a_1, b_1, \ldots, a_\ell, b_\ell)) = \exp(X_\mu).$$

More geometrically, given $(a_1, b_1, \ldots, a_\ell, b_\ell) \in V^{\ell,0}_{\mathrm{YM}}(G_{\mathbb{R}})_\mu$, let P be the underlying topological $L^I_{\mathbb{R}}$-bundle. Then $o(a_1, b_1, \ldots, a_\ell, b_\ell) = o_2(P)$ under the identification $\pi_1(L^I_{\mathbb{R}}) \cong H^2(\Sigma^\ell_0; \pi_1(L^I_{\mathbb{R}}))$. It is shown in [**HL3**] that for $\ell \geq 1$, $o^{-1}(k)$ is nonempty and connected for all $k \in p^{-1}_{L^I_{\mathbb{R}}}(\mu)$. We conclude that

PROPOSITION 4.2. *For any $I \subseteq \Delta$ and $\mu \in \Xi^I_+$, there is a bijection*

$$\pi_0\left(V^{\ell,0}_{\mathrm{YM}}(G_{\mathbb{R}})_\mu\right) \cong p^{-1}_{L^I_{\mathbb{R}}}(\mu).$$

Consider the short exact sequence of abelian groups:

$$\begin{array}{ccccccccc} 0 & \longrightarrow & \pi_1((L^I_{\mathbb{R}})_{ss}) & \stackrel{i}{\longrightarrow} & \pi_1(L^I_{\mathbb{R}}) & \stackrel{p_{L^I_{\mathbb{R}}}}{\longrightarrow} & \pi_1(L^I_{\mathbb{R}}/(L^I_{\mathbb{R}})_{ss}) & \longrightarrow & 0 \\ & & \| & & \| & & \| & & \\ & & \widehat{\Lambda}^{L^I}/\Lambda^{L^I} & & \pi_1(H)/\Lambda^{L^I} & & \pi_1(H)/\widehat{\Lambda}^{L^I} & & \end{array}$$

There is a bijection

$$\pi_0(V^{\ell,0}_{\mathrm{YM}}(G_{\mathbb{R}})_\mu / L^I_{\mathbb{R}}) \to p^{-1}_{L^I_{\mathbb{R}}}(\mu).$$

Given any $\beta \in p^{-1}_{L^I_{\mathbb{R}}}(\mu)$, there is a bijection

$$\pi_1((L^I_{\mathbb{R}})_{ss}) \to p^{-1}_{L^I_{\mathbb{R}}}(\mu), \quad \alpha \mapsto i(\alpha) + \beta.$$

4.3. Equivariant Poincaré series

Given a C^∞ principal G-bundle ξ_0 over Σ^ℓ_0, let

$$\Xi_{\xi_0} = \{\mu \in \bigcup_{I \subseteq \Delta} \Xi^I_+ \mid \mu \text{ is of Atiyah-Bott type for } \xi_0\}.$$

The Harder-Narasimhan stratification of the space $\mathcal{C}(\xi_0)$ of $(0,1)$-connections on ξ_0 is given by

$$\mathcal{C}(\xi_0) = \bigcup_{\mu \in \Xi_{\xi_0}} \mathcal{C}_\mu(\xi_0).$$

Recall that $\mathcal{C}(\xi_0)$ is an infinite dimensional complex affine space, and each strata $\mathcal{C}_\mu(\xi_0)$ is a complex submanifold of complex codimension

$$d_\mu = \sum_{\alpha(\mu) > 0, \alpha \in R^+} (\alpha(\mu) + \ell - 1) \tag{4.2}$$

Let P be a C^∞ principal $G_{\mathbb{R}}$-bundle over Σ_0^ℓ such that $P \times_{G_{\mathbb{R}}} G = \xi_0$, and let $\mathcal{A}(P)$ be the space of $G_{\mathbb{R}}$-connections on P. Then $\mathcal{A}(P) \cong \mathcal{C}(\xi_0)$ as infinite dimensional complex affine spaces. In [**AB**], Atiyah and Bott conjectured that the Morse stratification of the Yang-Mills functional on $\mathcal{A}(P)$ exists and coincides with the Harder-Narasimhan stratification on $\mathcal{C}(\xi_0)$ under the isomorphism $\mathcal{A}(P) \cong \mathcal{C}(\xi_0)$. The conjecture was proved by Daskalopoulos in [**Da**]. Atiyah and Bott showed that the Harder-Narasimhan stratification is $\mathcal{G}(\xi_0)$-perfect over $\mathbb{Q}$, where $\mathcal{G}(\xi_0) = \mathrm{Aut}(\xi_0)$ is the gauge group of ξ_0. Therefore,

$$P_t^{\mathcal{G}(\xi_0)}(\mathcal{C}(\xi_0);\mathbb{Q}) = \sum_{\mu\in\Xi_{\xi_0}} t^{2d_\mu} P_t^{\mathcal{G}(\xi_0)}(\mathcal{C}_\mu(\xi_0);\mathbb{Q}). \tag{4.3}$$

Let $\mathcal{A}_\mu(P) \subset \mathcal{A}(P)$ be the Morse stratum corresponding to $\mathcal{C}_\mu(\xi_0) \subset \mathcal{C}(\xi_0)$. It is the stable manifold of a connected component $\mathcal{N}_\mu(P)$ of $\mathcal{N}(P)$. Let $(G_{\mathbb{R}})_\mu = (G_{\mathbb{R}})_{X_\mu}$. Then μ and P uniquely determine a topological principal $(G_{\mathbb{R}})_\mu$-bundle P_μ. Let $X_{\mathrm{YM}}^{\ell,0}(G_{\mathbb{R}})_\mu^P$ denote the connected component of $X_{\mathrm{YM}}^{\ell,0}(G_{\mathbb{R}})_\mu$ which corresponds to $P \in \mathrm{Prin}_{G_{\mathbb{R}}}(\Sigma_0^\ell)$, and let $V_{\mathrm{YM}}^{\ell,0}(G_{\mathbb{R}})_\mu^{P_\mu}$ denote the connected component of $V_{\mathrm{YM}}^{\ell,0}(G_{\mathbb{R}})_\mu$ which corresponds to $P_\mu \in \mathrm{Prin}_{(G_{\mathbb{R}})_\mu}(\Sigma_0^\ell)$. Then $V_{\mathrm{YM}}^{\ell,0}(G_{\mathbb{R}})_\mu^{P_\mu}$ can be identified with the representation variety $V_{ss}(P_\mu)$ of *central* Yang-Mills connections on P_μ. We have homeomorphisms

$$\mathcal{N}_\mu(P)/\mathcal{G}(P) \cong X_{\mathrm{YM}}^{\ell,0}(G_{\mathbb{R}})_\mu^P/G_{\mathbb{R}} \cong V_{ss}(P_\mu)/(G_{\mathbb{R}})_\mu$$

and homotopy equivalences of homotopic orbit spaces:

$$\mathcal{N}_\mu(P)^{h\mathcal{G}(P)} \sim \left(X_{\mathrm{YM}}^{\ell,0}(G_{\mathbb{R}})_\mu^P\right)^{hG_{\mathbb{R}}} \sim V_{ss}(P_\mu)^{h(G_{\mathbb{R}})_\mu}.$$

Combined with the homotopy equivalence $\mathcal{C}_\mu(\xi_0)^{h\mathcal{G}(\xi_0)} \sim \mathcal{N}_\mu(P)^{h\mathcal{G}(P)}$, we conclude that

$$P_t^{\mathcal{G}(\xi_0)}(\mathcal{C}_\mu(\xi_0);\mathbb{Q}) = P_t^{G_{\mathbb{R}}}(X_{\mathrm{YM}}^{\ell,0}(G_{\mathbb{R}})_\mu^P;\mathbb{Q}) = P_t^{(G_{\mathbb{R}})_\mu}(V_{ss}(P_\mu);\mathbb{Q}).$$

REMARK 4.3. The connectedness of $\mathcal{N}_\mu(P)$ implies the connectedness of $V_{ss}(P_\mu)$, but not vise versa, because $\mathcal{G}_0(P)$ is not connected in general. We know $\mathcal{C}_\mu(\xi_0)$ is connected by results in [**AB**], and $\mathcal{N}_\mu(P) = \mathcal{N}(P) \cap \mathcal{A}_\mu(P)$ is a deformation retract of $\mathcal{A}_\mu(P) \cong \mathcal{C}_\mu(\xi_0)$ by results in [**Da, Rå**], so $\mathcal{N}_\mu(P)$ is connected.

Suppose that $\ell \geq 2$. Then there is a unique $\mu_0 \in \Xi_{\xi_0}$ such that $d_{\mu_0} = 0$. Then $\mathcal{C}_{\mu_0}(\xi_0) = \mathcal{C}_{ss}(\xi_0)$, the semi-stable stratum. Let

$$\mathcal{A}_{ss}(P) = \mathcal{A}_{\mu_0}(P), \quad \mathcal{N}_{ss}(P) = \mathcal{N}_{\mu_0}(P), \quad \Xi'_{\xi_0} = \Xi_{\xi_0} \setminus \{\mu_0\}.$$

Then

$$\mathcal{N}_{ss}(P)/\mathcal{G}_0(P) \cong V_{ss}(P).$$

The identity (4.3) can be rewritten as

$$P_t(B\mathcal{G}(P);\mathbb{Q}) = P_t^{G_{\mathbb{R}}}(V_{ss}(P);\mathbb{Q}) + \sum_{\mu\in\Xi'_{\xi_0}} t^{2d_\mu} P_t^{(G_{\mathbb{R}})_\mu}(V_{ss}(P_\mu);\mathbb{Q}) \tag{4.4}$$

where $P_t(B\mathcal{G}(P);\mathbb{Q})$ is given by Theorem 2.4. This allows one to compute

$$P_t^{G_{\mathbb{R}}}(V_{ss}(P);\mathbb{Q})$$

recursively.

When $G = GL(n,\mathbb{C})$, equivalent inductive procedure was derived by Harder and Narasimhan by number theoretic method in [**HN**]. Zagier provided an explicit closed formula which solves the recursion relation for $GL(n,\mathbb{C})$ [**Za**]. Laumon and Rapoport found an explicit closed formula which solves the recursion relation for general compact G. When G_{ss} is not simply connected, the recursion relation [**LR**, Theorem 3.2] that they solved is not exactly the recursion relation (4.4). The closed formula which solves (4.4) is the following slightly modified version of [**LR**, Theorem 3.4] (see Appendix A for details):

THEOREM 4.4. *Suppose that* $\xi_0 = P\times_{G_{\mathbb{R}}} G$ *and*

$$c_1(\xi_0)=\mu\in\pi_1(G)=\pi_1(H)/\Lambda.$$

Then

$$P_t^{G_{\mathbb{R}}}(V_{ss}(P)) =$$

$$\sum_{I\subseteq\Delta}(-1)^{\dim_{\mathbb{C}}\mathfrak{z}_{L^I}-\dim_{\mathbb{C}}\mathfrak{z}_G}P_t(B\mathcal{G}^{L^I_{\mathbb{R}}};\mathbb{Q})\frac{t^{2\dim_{\mathbb{C}}U^I(\ell-1)}}{\prod_{\alpha\in I}(1-t^{4\langle\rho_I,\alpha^\vee\rangle})}\cdot t^{4\sum_{\alpha\in I}\langle\rho_I,\alpha^\vee\rangle\langle\varpi_\alpha(\mu)\rangle}$$

where

$$\rho^I=\frac{1}{2}\sum_{\substack{\beta\in R_+\\ \langle\beta,\alpha^\vee\rangle>0\text{ for some }\alpha\in I}}\beta,$$

$\varpi_\alpha(\mu)\in\mathbb{Q}/\mathbb{Z}$, *and* $\langle x\rangle\in\mathbb{Q}$ *is the unique representative of the class* $x\in\mathbb{Q}/\mathbb{Z}$ *such that* $0<\langle x\rangle\leq 1$.

Theorem 4.4 coincides with [**LR**, Theorem 3.2] when G_{ss} is simply connected, for example, when $G_{\mathbb{R}}=U(n)$, $G=GL(n,\mathbb{C})$. When $G_{\mathbb{R}}=U(n)$, Theorem 4.4 specializes to the closed formula derived by Zagier in [**Za**] (see [**LR**, Section 4] for details):

THEOREM 4.5 ([**Za**], [**LR**, Section 4]).

$$\begin{aligned}&P_t^{U(n)}(X^{\ell,0}_{\mathrm{YM}}(U(n))_{\frac{k}{n},\dots,\frac{k}{n}})\\ =&\sum_{r=1}^{n}\sum_{\substack{n_1,\dots,n_r\in\mathbb{Z}_{>0}\\ \sum n_j=n}}(-1)^{r-1}\prod_{i=1}^{r}\frac{\prod_{j=1}^{n_i}(1+t^{2j-1})^{2\ell}}{(1-t^{2n_i})\prod_{j=1}^{n_i-1}(1-t^{2j})^2}\\ &\cdot\frac{t^{2(\ell-1)\sum_{i<j}n_in_j}}{\prod_{i=1}^{r-1}(1-t^{2(n_i+n_{i+1})})}\cdot t^{2\sum_{i=1}^{r-1}(n_i+n_{i+1})\langle(n_1+\cdots+n_i)(-\frac{k}{n})\rangle}\end{aligned}$$

REMARK 4.6. For $n\geq 2$, we have

$$\begin{aligned}&P_t^{U(n)}(X^{\ell,0}_{\mathrm{YM}}(U(n))_{0,\dots,0})=P_t^{U(n)}(X^{\ell,0}_{\text{flat}}(U(n)))\\ =&P_t^{U(1)}(X^{\ell,0}_{\text{flat}}(U(1)))P_t^{SU(n)}(X^{\ell,0}_{\text{flat}}(SU(n)))=\frac{(1+t)^{2\ell}}{1-t^2}P_t^{SU(n)}(X^{\ell,0}_{\text{flat}}(SU(n)))\end{aligned}$$

So Theorem 4.5 also gives a formula for $P_t^{SU(n)}(X^{\ell,0}_{\text{flat}}(SU(n)))$.

EXAMPLE 4.7.

$$\begin{aligned}&P_t^{U(2)}(X^{\ell,0}_{\mathrm{YM}}(U(2))_{\frac{k}{2},\frac{k}{2}})\\ =&\frac{(1+t)^{2\ell}(1+t^3)^{2\ell}}{(1-t^4)(1-t^2)^2}+(-1)\left(\frac{(1+t)^{2\ell}}{1-t^2}\right)^2\cdot\frac{t^{2(\ell-1)}}{1-t^4}t^{4\langle-\frac{k}{2}\rangle}\\ =&\frac{(1+t)^{2\ell}}{(1-t^2)^2(1-t^4)}\left((1+t^3)^{2\ell}-t^{2\ell-2+4\langle-\frac{k}{2}\rangle}(1+t)^{2\ell}\right)\end{aligned}$$

where

$$\langle -k/2\rangle = \begin{cases} 1 & k \text{ even} \\ 1/2 & k \text{ odd} \end{cases}$$

So

$$P_t^{U(2)}(X_{\mathrm{YM}}^{\ell,0}(U(2))_{\frac{k}{2},\frac{k}{2}}) = \begin{cases} \frac{(1+t)^{2\ell}}{(1-t^2)^2(1-t^4)}\left((1+t^3)^{2\ell} - t^{2\ell+2}(1+t)^{2\ell}\right) & k \text{ even} \\ \frac{(1+t)^{2\ell}}{(1-t^2)^2(1-t^4)}\left((1+t^3)^{2\ell} - t^{2\ell}(1+t)^{2\ell}\right) & k \text{ odd} \end{cases}$$

and

$$P_t^{SU(2)}(X_{\mathrm{flat}}^{\ell,0}(SU(2))) = \frac{(1+t^3)^{2\ell}}{(1-t^2)(1-t^4)} - \frac{t^{2\ell+2}(1+t)^{2\ell}}{(1-t^2)(1-t^4)}.$$

EXAMPLE 4.8.

$$\begin{aligned} &P_t^{SU(3)}(X_{\mathrm{flat}}^{\ell,0}(SU(3))) \\ = &\frac{(1+t^3)^{2\ell}(1+t^5)^{2\ell}}{(1-t^2)(1-t^4)^2(1-t^6)} - 2\frac{(1+t)^{2\ell}(1+t^3)^{2\ell}t^{4\ell+2}}{(1-t^2)^2(1-t^4)(1-t^6)} + \frac{(1+t)^{4\ell}t^{6\ell+2}}{(1-t^2)^2(1-t^4)^2} \end{aligned}$$

$$\begin{aligned} &P_t^{SU(4)}(X_{\mathrm{flat}}^{\ell,0}(SU(4))) \\ = &\frac{(1+t^3)^{2\ell}(1+t^5)^{2\ell}(1+t^7)^{2\ell}}{(1-t^2)(1-t^4)^2(1-t^6)^2(1-t^8)} - 2\frac{(1+t)^{2\ell}(1+t^3)^{2\ell}(1+t^5)^{2\ell}t^{6\ell+2}}{(1-t^2)^2(1-t^4)^2(1-t^6)(1-t^8)} \\ &- \frac{(1+t)^{2\ell}(1+t^3)^{4\ell}t^{8\ell}}{(1-t^2)^3(1-t^4)^2(1-t^8)} + 2\frac{(1+t)^{4\ell}(1+t^3)^{2\ell}t^{10\ell}}{(1-t^2)^3(1-t^4)^2(1-t^6)} \\ &+ \frac{(1+t)^{4\ell}(1+t^3)^{2\ell}t^{10\ell+2}}{(1-t^2)^3(1-t^4)(1-t^6)^2} - \frac{(1+t)^{6\ell}t^{12\ell}}{(1-t^2)^3(1-t^4)^3} \end{aligned}$$

We will use Theorem 4.4 to write down explicit closed formula for $SO(2n+1)$, $SO(2n)$, and $Sp(n)$ in Section 5.2, Section 6.2, and Section 7.2, respectively.

4.4. Involution on the Weyl Chamber

Let $\pi : \tilde{\Sigma} \to \Sigma$ be the orientable double cover of a closed, compact, connected, nonorientable surface Σ, and let $\tau : \tilde{\Sigma} \to \tilde{\Sigma}$ be the deck transformation. Let P be a principal $G_{\mathbb{R}}$-bundle over Σ, and let $\tilde{P} = \pi^* P$. Then $\tilde{P}$ and $\xi_0 = \tilde{P} \times_{G_{\mathbb{R}}} G$ are topologically trivial. There is an involution $\tilde{\tau}_s : \tilde{P} \to \tilde{P}$ which covers the anti-holomorphic involution $\tau : \tilde{\Sigma} \to \tilde{\Sigma}$. Under the trivialization $\tilde{P} \cong \tilde{\Sigma} \times G_{\mathbb{R}}$, τ_s is given by $(x, h) \mapsto (\tau(x), s(x)h)$, where $s : \tilde{\Sigma} \to G_{\mathbb{R}}$ satisfies $s(\tau(x)) = s(x)^{-1}$ (see [**HL4**, Section 3.2] for details).

Let $\mathcal{A}(P)$ and $\mathcal{A}(\tilde{P})$ denote the space of $G_{\mathbb{R}}$-connections on P and on $\tilde{P}$ respectively, and let $\mathcal{C}(\xi_0)$ be the space of $(0,1)$-connections on the principal G-bundle ξ_0. Then $\tilde{\tau}_s$ induces an involution $\tilde{\tau}_s^* : \mathcal{A}(\tilde{P}) \to \mathcal{A}(\tilde{P})$. Since $\tilde{P}$ and ξ_0 are topologically trivial, we may identify $\mathcal{A}(\tilde{P})$ with $\Omega^1(\tilde{\Sigma}, \mathfrak{g}_{\mathbb{R}})$ and identify $\mathcal{C}(\xi_0)$ with $\Omega^{0,1}(\tilde{\Sigma}, \mathfrak{g})$. Let $j : \Omega^1(\tilde{\Sigma}, \mathfrak{g}_{\mathbb{R}}) \to \Omega^{0,1}(\tilde{\Sigma}, \mathfrak{g})$ be defined as in the first paragraph of Chapter 3. Given $X = X_1 + \sqrt{-1}X_2 \in \mathfrak{g}$, where $X_1, X_2 \in \mathfrak{g}_{\mathbb{R}}$, define $\bar{X} = X_1 - \sqrt{-1}X_2$; given $X : \tilde{\Sigma} \to \mathfrak{g}$, define $\bar{X} : \tilde{\Sigma} \to \mathfrak{g}$ by $x \mapsto \overline{X(x)}$. Then $j \circ \tilde{\tau}_s^* \circ j^{-1} : \mathcal{C}(\xi_0) \to \mathcal{C}(\xi_0)$ is given by

$$X \otimes \theta \mapsto \mathrm{Ad}(s)\overline{\tau^* X} \otimes \overline{\tau^* \theta}$$

where $X \in \Omega^0(\tilde{\Sigma}, \mathfrak{g})$ and $\theta \in \Omega^{0,1}(\tilde{\Sigma})$. From now on, we denote $j \circ \tilde{\tau}_s^* \circ j^{-1}$ by $\tilde{\tau}_s^*$. We have isomorphisms of real affine spaces $\mathcal{A}(P) \cong \mathcal{A}(\tilde{P})^{\tilde{\tau}_s^*} \cong \mathcal{C}(\xi_0)^{\tilde{\tau}_s^*}$.

We will define an involution τ' on the positive Weyl chamber $\overline{C}_0$ such that $\tilde{\tau}_s^*\mathcal{C}_\mu = \mathcal{C}_{\tau'(\mu)}$, where $\mu \in \overline{C}_0$ is of Atiyah-Bott type for ξ_0 and $\mathcal{C}_\mu$ is the associated stratum in $\mathcal{C}(\xi_0)$.

The set

$$-C_0 = \{-Y \mid Y \in C_0\} \subset \sqrt{-1}\mathfrak{h}_{\mathbb{R}}.$$

is another Weyl chamber. There is a unique element w in the Weyl group W such that $w \cdot C_0 = -C_0$. We have $w^2 \cdot C_0 = C_0$, so $w^2 = \mathrm{id}_{\sqrt{-1}\mathfrak{h}_{\mathbb{R}}}$. Define $\tau' : \sqrt{-1}\mathfrak{h}_{\mathbb{R}} \to \sqrt{-1}\mathfrak{h}_{\mathbb{R}}$ by $X \mapsto w \cdot (-X)$. Recall that τ induces an involution on the symmetric representation variety which maps $X \in \mathfrak{g}_{\mathbb{R}}$ to $-\mathrm{Ad}(\bar{c})X \in \mathfrak{g}_{\mathbb{R}}$ (see [**HL4**, Section 4.5]). Given $Y \in \overline{C}_0$, $\tau'(Y)$ is the unique vector in $\overline{C}_0$ which is in the orbit $G \cdot (-Y) = G \cdot (-\mathrm{Ad}(\bar{c})(Y))$ of the adjoint action of G on $\mathfrak{g}$. Thus τ' is induced by the involution τ on the symmetric representation variety. To simplify notation, from now on we will write τ instead of τ'. Obviously $\tau(\overline{C}_0) = \overline{C}_0$. Given $Y \in \overline{C}_0$, $\tau(Y) = Y$ if and only if $Y \in \overline{C}_0$ is conjugate to $-Y$. In this case, we have $\mathrm{Ad}(\epsilon)Y = -Y$, where $\epsilon \in N(H_{\mathbb{R}}) \subset G_{\mathbb{R}}$ represents $w \in W = N(H_{\mathbb{R}})/H_{\mathbb{R}}$.

To demonstrate the above discussion, we list some examples of classical Lie groups.

EXAMPLE 4.9. Let $G_{\mathbb{R}} = U(n)$. then

$$\begin{aligned} \overline{C}_0 &= \{\mathrm{diag}(t_1, \dots, t_n) \mid t_1, \dots, t_n \in \mathbb{R}, t_1 \geq \cdots \geq t_n\} \\ -\overline{C}_0 &= \{\mathrm{diag}(v_1, \dots, v_n) \mid v_1, \dots, v_n \in \mathbb{R}, v_1 \leq \cdots \leq v_n\} \end{aligned}$$

There exists a unique w in $W \cong S(n)$, the symmetric group, such that $w(\overline{C}_0) = -\overline{C}_0$. In fact, $w \cdot \mathrm{diag}(t_1, \dots, t_n) = \mathrm{diag}(t_n, \dots, t_1)$ is the action of such w on $\sqrt{-1}\mathfrak{h}_{\mathbb{R}}$. Thus, the involution $\tau(Y)$ defined as $w\cdot(-Y)$ gives us $\tau(\mathrm{diag}(t_1, \dots, t_n)) = \mathrm{diag}(-t_n, \dots, -t_1)$, and Y is conjugate to $-Y$ (i.e. $\tau(Y) = Y$) if and only if $(t_1, \dots, t_n) = (-t_n, \dots, -t_1)$, or equivalently, if and only if Y is of the form $\mathrm{diag}(v_1, \dots, v_k, 0, \dots, 0, -v_k, \dots, -v_1)$.

EXAMPLE 4.10. Let $G_{\mathbb{R}} = SO(2n+1)$. then

$$\begin{aligned} \overline{C}_0 &= \{\sqrt{-1}\mathrm{diag}(t_1 J, \dots, t_n J, 0 I_1) \mid t_1 \geq \cdots \geq t_n \geq 0\}, \\ -\overline{C}_0 &= \{\sqrt{-1}\mathrm{diag}(v_1 J, \dots, v_n J, 0 I_1) \mid v_1 \leq \cdots \leq v_n \leq 0\}, \end{aligned}$$

where

$$J = \begin{pmatrix} 0 & -1 \\ 1 & 0 \end{pmatrix}.$$

The unique w in $W \cong G(n)$, the wreath product of $\mathbb{Z}_2$ by $S(n)$, that maps $\overline{C}_0$ to $-\overline{C}_0$, acts as $w \cdot \sqrt{-1}\mathrm{diag}(t_1 J, \dots, t_n J, 0 I_1) = \sqrt{-1}\mathrm{diag}(-t_1 J, \dots, -t_n J, 0 I_1)$. Thus $\tau : \sqrt{-1}\mathfrak{h}_{\mathbb{R}} \to \sqrt{-1}\mathfrak{h}_{\mathbb{R}}$ is the identity map. Any $Y \in \overline{C}_0$ is conjugate to the negative of itself. Let

$$H = \begin{pmatrix} 1 & 0 \\ 0 & -1 \end{pmatrix}$$

and let $H_n = \mathrm{diag}(\underbrace{H, \dots, H}_{n})$. The element

$$\epsilon = \mathrm{diag}(H_n, (-1)^n) \in SO(2n+1)$$

satisfies $\mathrm{Ad}(\epsilon)Y = -Y$ for all $Y \in \sqrt{-1}\mathfrak{h}_{\mathbb{R}}$ and $\epsilon^2 = e$.

EXAMPLE 4.11. Let $G_{\mathbb{R}} = SO(2n)$. Then

$$\begin{aligned} \overline{C}_0 &= \{\sqrt{-1}\mathrm{diag}(t_1 J, \dots, t_n J) \mid t_1 \geq \cdots \geq \mid t_n \mid \geq 0\}, \\ -\overline{C}_0 &= \{\sqrt{-1}\mathrm{diag}(v_1 J, \dots, v_n J) \mid v_1 \leq \cdots \leq - \mid v_n \mid \leq 0\}. \end{aligned}$$

The unique w in $W \cong SG(n)$, the subgroup of $G(n)$ consisting of even permutations, that maps $\overline{C}_0$ to $-\overline{C}_0$, belongs to the $\mathbb{Z}_2$ part of $SG(n)$, and

$$w \cdot \sqrt{-1}\mathrm{diag}(t_1 J, \dots, t_n J) = \sqrt{-1}\mathrm{diag}(-t_1 J, \dots, -t_{n-1} J, (-1)^{n-1} t_n J).$$

Thus

$$\tau\left(\sqrt{-1}\mathrm{diag}(t_1 J, \dots, t_n J)\right) = \sqrt{-1}\mathrm{diag}(t_1 J, \dots, t_{n-1} J, (-1)^n t_n J)$$

If n is even, then any $Y \in \overline{C}_0$ in conjugate to the negative of itself. If n is odd, then $Y \in \overline{C}_0$ is conjugate to $-Y$ iff Y is of the form $\sqrt{-1}\mathrm{diag}(t_1 J, \dots, t_{n-1} J, 0)$. Define

$$\epsilon = \begin{cases} H_n \\ \mathrm{diag}(H_{n-1}, I_2) \end{cases} \in SO(2n) \quad \text{if } n \text{ is } \begin{matrix} \text{even} \\ \text{odd} \end{matrix}$$

Then ϵ satisfies $\mathrm{Ad}(\epsilon)Y = -Y$ for all $Y \in \sqrt{-1}\mathfrak{h}_{\mathbb{R}}$ and $\epsilon^2 = e$.

EXAMPLE 4.12. Let $G_{\mathbb{R}} = Sp(n)$. Then

$$\begin{aligned} \overline{C}_0 &= \{\mathrm{diag}(t_1, \dots, t_n, -t_1, \dots, -t_n) \mid t_1 \geq \cdots \geq t_n \geq 0\}, \\ -\overline{C}_0 &= \{\mathrm{diag}(v_1, \dots, v_n, -v_1, \dots, -v_n) \mid v_1 \leq \cdots \leq v_n \leq 0\}. \end{aligned}$$

the unique w in $W \cong G(n)$, the wreath product of $\mathbb{Z}_2$ by $S(n)$, that maps $\overline{C}_0$ to $-\overline{C}_0$, acts as $w \cdot \mathrm{diag}(t_1, \dots, t_n, -t_1, \dots, -t_n) = \mathrm{diag}(-t_1, \dots, -t_n, t_1, \dots, t_n)$. Thus $\tau : \sqrt{-1}\mathfrak{h}_{\mathbb{R}} \to \sqrt{-1}\mathfrak{h}_{\mathbb{R}}$ is the identity map, and any $Y \in \overline{C}_0$ is conjugate to the negative of itself just as in the $SO(2n+1)$ case. The element

$$\epsilon = \begin{pmatrix} 0 & -I_n \\ I_n & 0 \end{pmatrix} \in Sp(n)$$

satisfies $Ad(\epsilon)Y = -Y$ for all $Y \in \sqrt{-1}\mathfrak{h}_{\mathbb{R}}$ but $\epsilon^2 \neq e$. Indeed, let $\tilde{\epsilon}$ be any element that satisfies $\mathrm{Ad}(\tilde{\epsilon})Y = -Y$ for all $Y \in \sqrt{-1}\mathfrak{h}_{\mathbb{R}}$. Then we must have $\tilde{\epsilon} = \epsilon u$ for some u in the maximal torus, and it is straightforward to check that $\tilde{\epsilon}^2 = -e$.

4.5. Connected components of the representation variety for nonorientable surfaces

$G_{\mathbb{R}}$ is connected, so the natural projection

$$X_{\mathrm{YM}}^{\ell,i}(G_{\mathbb{R}}) \to X_{\mathrm{YM}}^{\ell,i}(G_{\mathbb{R}})/G_{\mathbb{R}}$$

induces a bijection

$$\pi_0(X_{\mathrm{YM}}^{\ell,i}(G_{\mathbb{R}})) \to \pi_0(X_{\mathrm{YM}}^{\ell,i}(G_{\mathbb{R}}))/G_{\mathbb{R}}).$$

Any point in $X_{\mathrm{YM}}^{\ell,1}(G_{\mathbb{R}})/G_{\mathbb{R}}$ can be represented uniquely by

$$(a_1, b_1, \dots, a_\ell, b_\ell, c, X)$$

where $X \in \overline{C}_0$. Moreover, we must have $X \in \overline{C}_0^{\tau}$. Similarly, any point in $X_{\mathrm{YM}}^{\ell,2}(G_{\mathbb{R}})/G_{\mathbb{R}}$ can be represented uniquely by

$$(a_1, b_1, \dots, a_\ell, b_\ell, d, c, X)$$

where $X \in \overline{C}_0^{\tau}$.

Recall that τ is an $\mathbb{R}$-linear map from $\sqrt{-1}\mathfrak{h}_{\mathbb{R}}$ to $\sqrt{-1}\mathfrak{h}_{\mathbb{R}}$. Its dual τ^* is an $\mathbb{R}$-linear map from $(\sqrt{-1}\mathfrak{h}_{\mathbb{R}})^\vee$ to $(\sqrt{-1}\mathfrak{h}_{\mathbb{R}})^\vee$. This τ^* preserves Δ, the set of simple roots, and restricts to an involution on it. To simplify notation, we will also denote this involution by τ. Given $I \subseteq \Delta$ such that $\tau(I) = I$, let

$$(\Xi_+^I)^\tau = \{\mu \in \Xi_+^I \mid \tau(\mu) = \mu\}$$

Suppose that $(a_1, b_1, \ldots, a_\ell, b_\ell, c, X) \in X_{\mathrm{YM}}^{\ell,1}(G_{\mathbb{R}})$. Then there is a unique pair (μ, I), where $I \subseteq \Delta$, $\tau(I) = I$, and $\mu \in (\Xi_+^I)^\tau$, such that X is conjugate to $X_\mu = -2\pi\sqrt{-1}\mu$. Given $\mu \in (\Xi_+^I)^\tau$, where $I \subseteq \Delta$ and $\tau(I) = I$, define

$$\begin{aligned} X_{\mathrm{YM}}^{\ell,1}(G_{\mathbb{R}})_\mu &= \{(a_1, b_1, \ldots, a_\ell, b_\ell, c, X) \in G_{\mathbb{R}}{}^{2\ell+1} \times C_{\mu/2} \mid \\ &\qquad a_1, b_1, \ldots, a_\ell, b_\ell \in (G_{\mathbb{R}})_X, \mathrm{Ad}(c)X = -X, \prod_{i=1}^{\ell}[a_i, b_i] = \exp(X)c^2\} \end{aligned}$$

Where $C_{\mu/2}$ is the conjugacy class of $X_\mu/2$. We define $X_{\mathrm{YM}}^{\ell,2}(G_{\mathbb{R}})_\mu$ similarly. For $i = 1, 2$, $X_{\mathrm{YM}}^{\ell,i}(G_{\mathbb{R}})$ is a disjoint union of

$$\{X_{\mathrm{YM}}^{\ell,i}(G_{\mathbb{R}})_\mu \mid \mu \in (\Xi_+^I)^\tau, I \subseteq \Delta, \tau(I) = I\}.$$

When $G_{\mathbb{R}} = U(n)$, $\ell \geq 1$, each $X_{\mathrm{YM}}^{\ell,i}(G_{\mathbb{R}})_\mu$ is nonempty and has one or two connected components (see [**HL4**, Section 7]). We will see later that $X_{\mathrm{YM}}^{\ell,i}(G_{\mathbb{R}})_\mu$ can be empty for other classical groups (Section 5.3, Section 6.3, Section 6.4 and Section 7.3). When $X_{\mathrm{YM}}^{\ell,i}(G_{\mathbb{R}})_\mu$ is nonempty, it is a union of finitely many connected components of $X_{\mathrm{YM}}^{\ell,i}(G_{\mathbb{R}})$.

The reduction of $X_{\mathrm{YM}}^{\ell,i}(G_{\mathbb{R}})_\mu$ is more complicated because c is not in G_X. To do the reduction, we fix some $\epsilon \in G_{\mathbb{R}}$ such that the involution on $\overline{C}_0$ is given by $X \mapsto -\mathrm{Ad}(\epsilon)X$. Thus $\mathrm{Ad}(\epsilon)X = -X$ if X is fixed by the involution. For any $\mu \in (\Xi_+^I)^\tau$, where $\tau(I) = I$, we define *ϵ-reduced representation varieties*

$$V_{\mathrm{YM}}^{\ell,1}(G_{\mathbb{R}})_\mu = \{(a_1, b_1, \ldots, a_\ell, b_\ell, c') \in (L_{\mathbb{R}}^I)^{2\ell+1} \mid \prod_{i=1}^{\ell}[a_i, b_i] = \exp(\frac{X_\mu}{2})\epsilon c'\epsilon c'\} \tag{4.5}$$

(4.6)

$$V_{\mathrm{YM}}^{\ell,2}(G_{\mathbb{R}})_\mu = \{(a_1, b_1, \ldots, a_\ell, b_\ell, d, c') \in (L_{\mathbb{R}}^I)^{2\ell+2} \mid \prod_{i=1}^{\ell}[a_i, b_i] = \exp(\frac{X_\mu}{2})\epsilon c' d(\epsilon c')^{-1}d\}$$

For $i = 1, 2$, $L_{\mathbb{R}}^I$ acts on $V_{\mathrm{YM}}^{\ell,i}(G_{\mathbb{R}})_\mu$ by

$$g \cdot (c_1, \ldots, c_{2\ell+i}) = (gc_1g^{-1}, \ldots, gc_{2\ell+i-1}g^{-1}, \epsilon^{-1}g\epsilon c_{2\ell+i}g^{-1}).$$

Recall that $\mathrm{Ad}(\epsilon)(X_\mu) = -X_\mu$ and $L_{\mathbb{R}}^I = (G_{\mathbb{R}})_{X_\mu}$. So we have a homeomorphism

$$X_{\mathrm{YM}}^{\ell,i}(G_{\mathbb{R}})_\mu / G_{\mathbb{R}} \cong V_{\mathrm{YM}}^{\ell,i}(G_{\mathbb{R}})_\mu / L_{\mathbb{R}}^I$$

and a homotopy equivalence between homotopic orbit spaces:

$$X_{\mathrm{YM}}^{\ell,i}(G_{\mathbb{R}})_\mu{}^{hG_{\mathbb{R}}} \sim V_{\mathrm{YM}}^{\ell,i}(G_{\mathbb{R}})_\mu{}^{hL_{\mathbb{R}}^I}$$

When $G_{\mathbb{R}} = U(n)$, $V_{\mathrm{YM}}^{\ell,i}(G_{\mathbb{R}})_\mu$ can be viewed as a product of representation varieties for $U(m)$ $(m < n)$ of Σ_i^ℓ and of its double cover $\Sigma_0^{2\ell+i-1}$ (see [**HL4**, Section 7]. This is not the case for other classical groups. We will see in Section 5.3, Section 6.3, Section 6.4, and Section 7.3 that when $G_{\mathbb{R}} = SO(n)$ or $Sp(n)$,

$V^{\ell,i}_{\mathrm{YM}}(G_{\mathbb{R}})_\mu$ is a product of *twisted representation varieties* defined in Section 4.6 and Section 4.7 below.

4.6. Twisted representation varieties: $U(n)$

Given $n, k \in \mathbb{Z}$, $n > 0$, define *twisted representation varieties*

$$\tilde{V}^{\ell,1}_{n,k} = \Big\{(a_1, b_1, \dots, a_\ell, b_\ell, c) \in U(n)^{2\ell+1} \mid \prod_{i=1}^{\ell}[a_i, b_i] = e^{-2\pi\sqrt{-1}k/n} I_n \bar{c}c\Big\} \tag{4.7}$$

(4.8)

$$\tilde{V}^{\ell,2}_{n,k} = \Big\{(a_1, b_1, \dots, a_\ell, b_\ell, d, c) \in U(n)^{2\ell+2} \mid \prod_{i=1}^{\ell}[a_i, b_i] = e^{-2\pi\sqrt{-1}k/n} I_n \bar{c}\bar{d}\bar{c}^{-1}d\}$$

where $\bar{c}$ is the complex conjugate of c. In particular,

$$\tilde{V}^{\ell,1}_{1,k} = U(1)^{2\ell+1}, \quad \tilde{V}^{\ell,2}_{1,k} = U(1)^{2\ell+2}.$$

For $i = 1, 2$, $U(n)$ acts on $\tilde{V}^{\ell,i}_{n,k}$ by

$$g \cdot (a_1, b_1, \dots, a_\ell, b_\ell, c) = (ga_1g^{-1}, gb_1g^{-1}, \dots, ga_\ell g^{-1}, gb_\ell g^{-1}, \bar{g}cg^{-1}) \tag{4.9}$$

(4.10)

$$g \cdot (a_1, b_1, \dots, a_\ell, b_\ell, d, c) = (ga_1g^{-1}, gb_1g^{-1}, \dots, ga_\ell g^{-1}, gb_\ell g^{-1}, gdg^{-1}, \bar{g}cg^{-1})$$

We will show that

PROPOSITION 4.13. *$\tilde{V}^{\ell,i}_{n,k}$ is nonempty and connected for $\ell \geq 2i$.*

PROOF FOR $i = 1$. For any $(a_1, b_1, \dots, a_\ell, b_\ell, c) \in \tilde{V}^{\ell,1}_{n,k}$, we have

$$\det(a_i) = e^{\sqrt{-1}\theta_i}, \quad \det(b_i) = e^{\sqrt{-1}\phi_i}, \quad \det(c) = e^{\sqrt{-1}\theta}.$$

Define $\beta : [0, 1] \to U(n)^{2\ell+1}$ by

$$\beta(t) = (e^{-\sqrt{-1}t\theta_1/n}a_1, e^{-\sqrt{-1}t\phi_1/n}b_1, \dots, e^{-\sqrt{-1}t\theta_\ell/n}a_\ell, e^{-\sqrt{-1}t\phi_\ell/n}b_\ell, e^{-\sqrt{-1}t\theta/n}c).$$

Then the image of β lies in $\tilde{V}^{\ell,1}_{n,k}$, $\beta(0) = (a_1, b_1, \dots, a_\ell, b_\ell, c)$, and

$$\beta(1) \in W^{\ell,1}_{n,k} \stackrel{\text{def}}{=} \Big\{(a_1, b_1, \dots, a_\ell, b_\ell, c) \in SU(n)^{2\ell+1} | \prod_{i=1}^{\ell}[a_i, b_i] = e^{-2\pi\sqrt{-1}k/n} I_n \bar{c}c\Big\} \subset \tilde{V}^{\ell,1}_{n,k}.$$

So it suffices to show that $W^{\ell,1}_{n,k}$ is nonempty and connected.

Define $\pi : W^{\ell,1}_{n,k} \to SU(n)$ by $(a_1, b_1, \dots, a_\ell, b_\ell, c) \mapsto c$. Then $\pi^{-1}(c)$ is nonempty and connected for any $c \in SU(n)$. It remains to show that for any $c \in SU(n)$, there is a path $\gamma : [0, 1] \to W^{\ell,1}_{n,k}$ such that $\gamma(0) \in \pi^{-1}(e)$ and $\gamma(1) \in \pi^{-1}(c)$.

Let T be the maximal torus which consists of diagonal matrices in $SU(n)$. For any $c \in SU(n)$, there exist $g \in SU(n)$ such that $g^{-1}cg \in T$. We have

$$c = g \exp \xi g^{-1}, \quad \bar{c} = \bar{g} \exp(-\xi) \bar{g}^{-1}$$

for some $\xi \in \mathfrak{t}$. Let

$$\xi_0 = -2\pi\sqrt{-1}\frac{k}{n}\mathrm{diag}(I_{n-1}, (1-n)I_1) \in \mathfrak{t}.$$

Then $\exp(\xi_0) = e^{-2\pi\sqrt{-1}k/n} I_n$. Let ω be the coxeter element and a be the corresponding element in $SU(n)$. There are $\eta_0, \eta \in \mathfrak{t}$ such that

$$\omega \cdot \eta_0 - \eta_0 = \xi_0, \quad \omega \cdot \eta - \eta = \xi.$$

Let $a \in N(T)$ represent $\omega \in W = N(T)/T$. Then

$$\begin{aligned} a\exp(\eta_0 - t\eta)a^{-1}\exp(-\eta_0 + t\eta) &= \exp(\omega \cdot (\eta_0 - t\eta) - (\eta_0 - t\eta)) \\ &= \exp(\xi_0 - t\xi) \\ &= e^{-2\pi\sqrt{-1}k/n}\exp(-t\xi) \\ a\exp(t\eta)a^{-1}\exp(-t\eta) &= \exp(\omega \cdot (t\eta) - t\eta) = \exp(t\xi). \end{aligned}$$

Now since $SU(n)$ is connected, there are paths $\tilde{g} : [0,1] \to SU(n)$ such that $\tilde{g}(0) = e$ and $\tilde{g}(1) = g$. Now define $\gamma : [0,1] \to SU(n)^{2\ell+1}$ by

$$\gamma(t) = (a_1(t), b_1(t), a_2(t), b_2(t), e, \dots, e, c(t))$$

where

$$\begin{aligned} &a_1(t) = \overline{\tilde{g}(t)}a\left(\overline{\tilde{g}(t)}\right)^{-1}, \quad b_1(t) = \overline{\tilde{g}(t)}\exp(\eta_0 - t\eta)\left(\overline{\tilde{g}(t)}\right)^{-1} \\ &a_2(t) = \tilde{g}(t)a\tilde{g}(t)^{-1}, \quad b_2(t) = \tilde{g}(t)\exp(t\eta)\tilde{g}(t)^{-1} \\ &c(t) = \tilde{g}(t)\exp(t\xi)\tilde{g}(t)^{-1}. \end{aligned}$$

Then

$$[a_1(t), b_1(t)] = e^{-2\pi\sqrt{-1}k/n}\overline{c(t)}, \quad [a_2(t), b_2(t)] = c(t),$$

so the image of γ lies in $W^{\ell,1}_{n,k}$. We have

$$\begin{aligned} \gamma(0) &= (a, \exp(\eta_0), a, e, e, \dots, e, e, e) \in \pi^{-1}(e) \\ \gamma(1) &= (\bar{g}a\bar{g}^{-1}, \bar{g}\exp(\eta_0 - \eta)\bar{g}^{-1}, gag^{-1}, g\exp(\eta)\bar{g}^{-1}, e, \cdots, e, c) \in \pi^{-1}(c). \end{aligned}$$

□

PROOF FOR $i = 2$. For any $(a_1, b_1, \dots, a_\ell, b_\ell, d, c) \in \tilde{V}^{\ell,2}_{n,k}$, we have

$$\det(a_i) = e^{\sqrt{-1}\theta_i}, \quad \det(b_i) = e^{\sqrt{-1}\phi_i}, \quad \det(c) = e^{\sqrt{-1}\theta}, \quad \det(d) = e^{\sqrt{-1}\phi}.$$

Define $\beta : [0,1] \to U(n)^{2\ell+2}$ by

$$\begin{aligned} \beta(t) = (&e^{-\sqrt{-1}t\theta_1/n}a_1, e^{-\sqrt{-1}t\phi_1/n}b_1, \dots, \\ &e^{-\sqrt{-1}t\theta_\ell/n}a_\ell, e^{-\sqrt{-1}t\phi_\ell/n}b_\ell, e^{-\sqrt{-1}t\phi/n}d, e^{-\sqrt{-1}t\theta/n}c). \end{aligned}$$

Then the image of β lies in $\tilde{V}^{\ell,2}_{n,k}$, $\beta(0) = (a_1, b_1, \dots, a_\ell, b_\ell, d, c)$, and

$$\beta(1) \in W^{\ell,2}_{n,k} \stackrel{\text{def}}{=} \Big\{(a_1, b_1, \dots, a_\ell, b_\ell, d, c) \in SU(n)^{2\ell+2} \,\Big|\, \prod_{i=1}^{\ell}[a_i, b_i] = e^{-2\pi\sqrt{-1}k/n} I_n \bar{c}\bar{d}c^{-1}d\Big\} \subset \tilde{V}^{\ell,2}_{n,k}.$$

So it suffices to show that $W^{\ell,2}_{n,k}$ is nonempty and connected.

Define $\pi : W^{\ell,2}_{n,k} \to SU(n)^2$ by $(a_1, b_1, \dots, a_\ell, b_\ell, d, c) \mapsto (d, c)$. Then $\pi^{-1}(d,c)$ is nonempty and connected for any $(d,c) \in SU(n)^2$. It remains to show that for any $(d,c) \in SU(n)^2$, there is a path $\gamma : [0,1] \to W^{\ell,2}_{n,k}$ such that $\gamma(0) \in \pi^{-1}(e,e)$ and $\gamma(1) \in \pi^{-1}(d,c)$.

Let T be the maximal torus which consists of diagonal matrices in $SU(n)$. For any $c, d \in SU(n)$, there exist $g_1, g_2 \in SU(n)$ such that $g_1^{-1}cg_1, g_2^{-1}dg_2 \in T$. We have

$$c = g_1 \exp \xi_1 g_1^{-1}, \quad \bar{c} = \bar{g}_1 \exp(-\xi_1)\bar{g}_1^{-1}, \quad d = g_2 \exp \xi_2 g_2^{-1}, \quad \bar{d} = \bar{g}_2 \exp(-\xi_2)\bar{g}_2^{-1}$$

for some $\xi_1, \xi_2 \in \mathfrak{t}$. Let

$$\xi_0 = -2\pi\sqrt{-1}\frac{k}{n}\mathrm{diag}(I_{n-1}, (1-n)I_1) \in \mathfrak{t}.$$

Then $\exp(\xi_0) = e^{-2\pi\sqrt{-1}k/n} I_n$. Let ω be the coxeter element and a be the corresponding element in $SU(n)$. There are $\eta_0, \eta_1, \eta_2 \in \mathfrak{t}$ such that

$$\omega \cdot \eta_j - \eta_j = \xi_j, \quad j = 0, 1, 2.$$

Let $a \in N(T)$ represent $\omega \in W = N(T)/T$. Then

$$\begin{aligned}
a \exp(\eta_0 - t\eta_1) a^{-1} \exp(-\eta_0 + t\eta_1) &= \exp(\omega \cdot (\eta_0 - t\eta_1) - (\eta_0 - t\eta_1)) \\
&= \exp(\xi_0 - t\xi_1) \\
&= e^{-2\pi\sqrt{-1}k/n} \exp(-t\xi_1) \\
a \exp(-t\eta_1) a^{-1} \exp(t\eta_1) &= \exp(\omega \cdot (-t\eta_1) + t\eta_1) = \exp(-t\xi_1) \\
a \exp(-t\eta_2) a^{-1} \exp(t\eta_2) &= \exp(\omega \cdot (-t\eta_2) + t\eta_2) = \exp(-t\xi_2).
\end{aligned}$$

Now since $SU(n)$ is connected, there are paths $\tilde{g}_1, \tilde{g}_2 : [0,1] \to SU(n)$ such that $\tilde{g}_j(0) = e$ and $\tilde{g}_j(1) = g_j$ for $j = 1, 2$. Now define $\gamma : [0,1] \to SU(n)^{2\ell+2}$ by

$$\gamma(t) = (a_1(t), b_1(t), a_2(t), b_2(t), a_3(t), b_3(t), a_4(t), b_4(t), e, \ldots, e, d(t), c(t))$$

where

$$\begin{aligned}
&a_1(t) = \overline{\tilde{g}_1(t)} a \left(\overline{\tilde{g}_1(t)}\right)^{-1}, \quad b_1(t) = \overline{\tilde{g}_1(t)} \exp(\eta_0 - t\eta_1) \left(\overline{\tilde{g}_1(t)}\right)^{-1} \\
&a_2(t) = \overline{\tilde{g}_2(t)} a \left(\overline{\tilde{g}_2(t)}\right)^{-1}, \quad b_2(t) = \overline{\tilde{g}_2(t)} \exp(-t\eta_2) \left(\overline{\tilde{g}_2(t)}\right)^{-1} \\
&a_3(t) = \overline{\tilde{g}_1(t)} a \left(\overline{\tilde{g}_1(t)}\right)^{-1}, \quad b_3(t) = \overline{\tilde{g}_1(t)} \exp(t\eta_1) \left(\overline{\tilde{g}_1(t)}\right)^{-1} \\
&a_4(t) = \tilde{g}_2(t) a \left(\tilde{g}_2(t)\right)^{-1}, \quad b_4(t) = \tilde{g}_2(t) \exp(t\eta_2) \left(\tilde{g}_2(t)\right)^{-1} \\
&c(t) = \tilde{g}_1(t) \exp(t\xi_1) \tilde{g}_1(t)^{-1}, \quad d(t) = \tilde{g}_2(t) \exp(t\xi_2) \tilde{g}_2(t)^{-1}
\end{aligned}$$

Then

$$\begin{aligned}
&[a_1(t), b_1(t)] = e^{-2\pi\sqrt{-1}k/n}\overline{c(t)}, \quad [a_2(t), b_2(t)] = \overline{d(t)}, \\
&[a_3(t), b_3(t)] = \overline{c(t)}^{-1}, \quad [a_4(t), b_4(t)] = d(t).
\end{aligned}$$

so the image of γ lies in $W^{\ell,2}_{n,k}$. We have

$$\begin{aligned}
\gamma(0) &= (a, \exp(\eta_0), a, e, a, e, a, e, e, \ldots, e, e, e) \in \pi^{-1}(e, e) \\
\gamma(1) &= (\overline{g_1} a \overline{g_1}^{-1}, \overline{g_1} \exp(\eta_0 - \eta_1) \overline{g_1}^{-1}, \overline{g_2} a \overline{g_2}^{-1}, \overline{g_2} \exp(-\eta_2) \overline{g_2}^{-1}, \\
&\quad \overline{g_1} a \overline{g_1}^{-1}, \overline{g_1} \exp(\eta_1) \overline{g_1}^{-1}, g_2 a g_2^{-1}, g_2 \exp(\eta_2) g_2^{-1}, e, \ldots, e, d, c) \in \pi^{-1}(d, c).
\end{aligned}$$

□

4.7. Twisted representation varieties: $SO(n)$

Let

$$O(n)_{\pm} = \{A \in O(n) \mid \det(A) = \pm 1\}.$$

Then $O(n)_+$ and $O(n)_-$ are the two connected components of $O(n)$, where $O(n)_+ = SO(n)$. For $n \geq 2$, define

$$V^{\ell,1}_{O(n),\pm 1} = \{(a_1, b_1, \dots, a_\ell, b_\ell, c) \in SO(n)^{2\ell} \times O(n)_{\pm} \mid \prod_{i=1}^{\ell} [a_i, b_i] = c^2\} \tag{4.11}$$

$$V^{\ell,2}_{O(n),\pm 1} = \{(a_1, b_1, \dots, a_\ell, b_\ell, d, c) \in SO(n)^{2\ell+1} \times O(n)_{\pm} \mid \prod_{i=1}^{\ell} [a_i, b_i] = cdc^{-1}d\} \tag{4.12}$$

Note that $V^{\ell,i}_{O(n),+1} = X^{\ell,i}_{\text{flat}}(SO(n))$. Recall that $X^{\ell,i}_{\text{flat}}(SO(n))$ has two connected components $X^{\ell,1}_{\text{flat}}(SO(n))^{+1}$ and $X^{\ell,2}_{\text{flat}}(SO(n))^{-1}$.

For $i = 1, 2$, $SO(n)$ acts on $V^{\ell,i}_{O(n),\pm 1}$ by

$$g \cdot (a_1, b_1, \dots, a_\ell, b_\ell, c) = (ga_1g^{-1}, gb_1g^{-1}, \dots, ga_\ell g^{-1}, gb_\ell g^{-1}, gcg^{-1}) \tag{4.13}$$

$$g \cdot (a_1, b_1, \dots, a_\ell, b_\ell, d, c) = (ga_1g^{-1}, gb_1g^{-1}, \dots, ga_\ell g^{-1}, gb_\ell g^{-1}, gdg^{-1}, gcg^{-1}) \tag{4.14}$$

When $n = 2$, we have diffeomorphisms $O(2)_+ \cong O(2)_- \cong U(1)$, and diffeomorphisms

$$V^{\ell,i}_{O(2),+1} \cong X^{\ell,i}_{\text{flat}}(U(1)) \cong U(1)^{2\ell+i-1} \times \{\pm 1\}$$

where $i = 1, 2$. For any $d \in SO(2)$ and $c \in O(2)_-$, we have

$$c^2 = I_2, \quad cdc^{-1}d = I_2,$$

so

$$\begin{aligned} V^{\ell,1}_{O(2),-1} &= \{(a_1, b_1, \dots, a_\ell, b_\ell, c) \in SO(2)^{2\ell} \times O(2)_- \mid I_2 = c^2\} \\ &= SO(2)^{2\ell} \times O(2)_-, \\ V^{\ell,2}_{O(2),-1} &= \{(a_1, b_1, \dots, a_\ell, b_\ell, d, c) \in SO(2)^{2\ell+1} \times O(2)_- \mid I_2 = cdc^{-1}d\} \\ &= SO(2)^{2\ell+1} \times O(2)_-. \end{aligned}$$

For $i = 1, 2$, $V^{\ell,i}_{O(2),-1}$ is diffeomorphic to $U(1)^{2\ell+i}$, thus nonempty and connected.

From now on, we assume that $n \geq 3$ so that $SO(n)$ is semisimple. Let $\rho : Pin(n) \to O(n)$ be the double cover defined in [**BD**, Chapter I, Section 6], and let $Pin(n)_{\pm} = \rho^{-1}(O(n)_{\pm})$. Then $Pin(n)_+$ and $Pin(n)_-$ are the two connected components of $Pin(n)$, where $Pin(n)_+ = Spin(n)$. Note that $Pin(n)_-$ is not a group because if $x, y \in Pin(n)_-$ then $xy \in Pin(n)_+$.

Recall that there is an obstruction map

$$o_2 : V^{\ell,1}_{O(n),+1} = X^{\ell,1}_{\text{flat}}(SO(n)) \to \operatorname{Ker}(\rho) = \{1, -1\} \subset Spin(n)$$

given by

$$(a_1, b_1, \dots, a_\ell, b_\ell, c) \mapsto \prod_{i=1}^{\ell} [\tilde{a}_i, \tilde{b}_i] \tilde{c}^{-2}$$

where $(\tilde{a}_1, \tilde{b}_1, \ldots, \tilde{a}_\ell, \tilde{b}_\ell, \tilde{c})$ is the preimage of $(a_1, b_1, \ldots, a_\ell, b_\ell, c)$ under $\rho^{2\ell+1}$: $Spin(n)^{2\ell+1} \to SO(n)^{2\ell+1}$. It is easy to check that o_2 does not depend on the choice of the liftings $(\tilde{a}_1, \tilde{b}_1, \ldots, \tilde{a}_\ell, \tilde{b}_\ell, \tilde{c})$ because $2\mathrm{Ker}(\rho) = \{1\}$. Similarly, there is an obstruction map $o_2 : V^{\ell,2}_{O(n),+1} = X^{\ell,2}_{\mathrm{flat}}(SO(n)) \to \{1, -1\}$ given by

$$(a_1, b_1, \ldots, a_\ell, b_\ell, d, c) \mapsto \prod_{i=1}^{\ell} [\tilde{a}_i, \tilde{b}_i](\tilde{c}\tilde{d}\tilde{c}^{-1}\tilde{d})^{-1}$$

where $(\tilde{a}_1, \tilde{b}_1, \ldots, \tilde{a}_\ell, \tilde{b}_\ell, \tilde{d}, \tilde{c})$ is the preimage of $(a_1, b_1, \ldots, a_\ell, b_\ell, d, c)$ under $\rho^{2\ell+2}$: $Spin(n)^{2\ell+2} \to SO(n)^{2\ell+2}$. Again, o_2 does not depend on the choice of $\tilde{a}_i$, $\tilde{b}_i$, $\tilde{d}$, $\tilde{c}$.

For $i = 1, 2$, define $V^{\ell,i,\pm 1}_{O(n),+1} = X^{\ell,i}_{\mathrm{flat}}(SO(n))^{\pm 1} = o_2^{-1}(\pm 1)$. Then $V^{\ell,i,+1}_{O(n),+1} = X^{\ell,i}_{\mathrm{flat}}(SO(n))^{+1}$ corresponds to flat connections on the trivial $SO(n)$-bundle ($w_2 = 0 \in H^2(\Sigma^\ell_i; \mathbb{Z}/2\mathbb{Z}) \cong \mathbb{Z}/2\mathbb{Z}$), while $V^{\ell,i,-1}_{O(n),+1} = X^{\ell,i}_{\mathrm{flat}}(SO(n))^{-1}$ corresponds to flat connections on the nontrivial $SO(n)$-bundle ($w_2 = 1 \in H^2(\Sigma^\ell_i; \mathbb{Z}/2\mathbb{Z}) \cong \mathbb{Z}/2\mathbb{Z}$). It was proved in [**HL2**] that $X^{\ell,i}_{\mathrm{flat}}(SO(n))^{+1}$ and $X^{\ell,i}_{\mathrm{flat}}(SO(n))^{-1}$ are nonempty and connected if $\ell \geq i$, i.e., $(\ell, i) \neq (0,1), (0,2), (1,2)$. The result is extended to the case $(1,2)$ in [**HL4**].

We now extend the definition of o_2 to $V^{\ell,i}_{O(n),-1}$. Define $o_2 : V^{\ell,1}_{O(n),-1} \to \{1, -1\} \subset Spin(n)$ by

$$(a_1, b_1, \ldots, a_\ell, b_\ell, c) \mapsto \prod_{i=1}^{\ell} [\tilde{a}_i, \tilde{b}_i]\tilde{c}^{-2}$$

where $(\tilde{a}_1, \tilde{b}_1, \ldots, \tilde{a}_\ell, \tilde{b}_\ell, \tilde{c})$ is the preimage of $(a_1, b_1, \ldots, a_\ell, b_\ell, c)$ under $\rho^{2\ell+1}$: $Spin(n)^{2\ell} \times Pin(n)_- \to SO(n)^{2\ell} \times O(n)_-$. It is easy to check that o_2 does not depend on the choice of $(\tilde{a}_1, \tilde{b}_1, \ldots, \tilde{a}_\ell, \tilde{b}_\ell, \tilde{c})$. Similarly, define $o_2 : V^{\ell,2}_{O(n),-1} \to \{1, -1\} \subset Spin(n)$ by

$$(a_1, b_1, \ldots, a_\ell, b_\ell, d, c) \mapsto \prod_{i=1}^{\ell} [\tilde{a}_i, \tilde{b}_i](\tilde{c}\tilde{d}\tilde{c}^{-1}\tilde{d})^{-1}$$

where $(\tilde{a}_1, \tilde{b}_1, \ldots, \tilde{a}_\ell, \tilde{b}_\ell, \tilde{d}, \tilde{c})$ is the preimage of $(a_1, b_1, \ldots, a_\ell, b_\ell, d, c)$ under $\rho^{2\ell+2}$: $Spin(n)^{2\ell+1} \times Pin(n)_- \to SO(n)^{2\ell+1} \times O(n)_-$. Again, o_2 does not depend on the choice of $(\tilde{a}_1, \tilde{b}_1, \ldots, \tilde{a}_\ell, \tilde{b}_\ell, \tilde{d}, \tilde{c})$. Define $V^{\ell,i,\pm 1}_{O(n),-1} = o_2^{-1}(\pm 1)$. We will show that

PROPOSITION 4.14. *Suppose that* $\ell \geq 2i$, *where* $i = 1, 2$, *and* $n \geq 3$. *Then* $V^{\ell,i,+1}_{O(n),-1}$ *and* $V^{\ell,i,-1}_{O(n),-1}$ *are nonempty and connected.*

PROOF. Define

$$\begin{aligned}
\tilde{V}^{\ell,1,\pm 1}_{Pin(n)_-} &= \{(\tilde{a}_1, \tilde{b}_1, \ldots, \tilde{a}_\ell, \tilde{b}_\ell, \tilde{c}) \in Spin(n)^{2\ell} \times Pin(n)_- \mid \prod_{i=1}^{\ell} [\tilde{a}_i, \tilde{b}_i]\tilde{c}^{-2} = \pm 1\} \\
\tilde{V}^{\ell,2,\pm 1}_{Pin(n)_-} &= \{(\tilde{a}_1, \tilde{b}_1, \ldots, \tilde{a}_\ell, \tilde{b}_\ell, \tilde{d}, \tilde{c}) \in Spin(n)^{2\ell+1} \times Pin(n)_- \mid \\
&\qquad \prod_{i=1}^{\ell} [\tilde{a}_i, \tilde{b}_i](\tilde{c}\tilde{d}\tilde{c}^{-1}\tilde{d})^{-1} = \pm 1\}
\end{aligned}$$

Then $\rho^{2\ell+i} : Spin(n)^{2\ell+i-1} \times Pin(n)_- \to SO(n)^{2\ell+i-1} \times O(n)_-$ restricts to a covering map $\tilde{V}^{\ell,i,\pm 1}_{Pin(n)_-} \to V^{\ell,i,\pm 1}_{O(n),-1}$. It suffices to prove that $\tilde{V}^{\ell,i,+1}_{Pin(n)_-}$ and $\tilde{V}^{\ell,i,-1}_{Pin(n)_-}$ are nonempty and connected for $\ell \geq 2i$.

$i = 1$. Define $\pi_\pm : \tilde{V}^{\ell,1,\pm 1}_{Pin(n)_-} \to Pin(n)_-$ by $(\tilde{a}_1, \tilde{b}_1, \ldots, \tilde{a}_\ell, \tilde{b}_\ell, \tilde{c}) \mapsto \tilde{c}$. Note that $Spin(n)$ is simply connected and $\tilde{c}^2, -\tilde{c}^2 \in Spin(n)$, so $\pi_\pm^{-1}(\tilde{c})$ is nonempty and connected for any $\tilde{c} \in Pin(n)_-$. Let $\epsilon_+ = e_1e_2e_3$, and let $\epsilon_- = e_1$. Then $\epsilon_+, \epsilon_- \in Pin(n)_-$, and $(\epsilon_\pm)^2 = \pm 1$. It suffices to show that for any $\tilde{c} \in Pin(n)_-$, there is a path $\gamma_\pm : [0,1] \to \tilde{V}^{\ell,1,\pm 1}_{Pin(n)_-}$ such that $\gamma_\pm(0) \in \pi_\pm^{-1}(\epsilon_\pm)$ and $\gamma_\pm(1) \in \pi_\pm^{-1}(\tilde{c})$.

Let T be the maximal torus of $Spin(n)$, and let $\mathfrak{t}$ be the Lie algebra of T. For any $\tilde{c} \in Pin(n)_-$, we have $(\epsilon_\pm)^{-1}\tilde{c} \in Spin(n)$, so there exists $g_\pm \in Spin(n)$ such that $(g_\pm)^{-1}(\epsilon_\pm)^{-1}\tilde{c}g_\pm \in T$. We have

$$\tilde{c} = \epsilon_\pm g_\pm \exp(\xi_\pm)(g_\pm)^{-1}$$

for some $\xi_+, \xi_- \in \mathfrak{t}$. Let ω be the coxeter element. There are $\eta_+, \eta_- \in \mathfrak{t}$ such that

$$\omega \cdot \eta_\pm - \eta_\pm = \xi_\pm.$$

Let $a \in N(T) \subset Spin(n)$ be the corresponding element which represents $\omega \in W = N(T)/T$. Then

$$a \exp(t\eta_\pm)a^{-1} \exp(-t\eta_\pm) = \exp(\omega \cdot t\eta_\pm - t\eta_\pm) = \exp(t\xi_\pm).$$

Now since $Spin(n)$ is connected, there are paths $\tilde{g}_\pm : [0,1] \to Spin(n)$ such that $\tilde{g}_\pm(0) = 1$ and $\tilde{g}_\pm(1) = g_\pm$. Now define $\gamma : [0,1] \to Spin(n)^{2\ell} \times Pin(n)_-$ by

$$\gamma_\pm(t) = (a_1^\pm(t), b_1^\pm(t), a_2^\pm(t), b_2^\pm(t), 1, \ldots, 1, c^\pm(t))$$

where

$$\begin{aligned}
&a_1^\pm(t) = \epsilon_\pm \tilde{g}_\pm(t) a(\epsilon_\pm \tilde{g}_\pm(t))^{-1}, \quad b_1^\pm(t) = \epsilon_\pm \tilde{g}_\pm(t) \exp(t\eta_\pm)(\epsilon_\pm \tilde{g}_\pm(t))^{-1},\\
&a_2^\pm(t) = \tilde{g}_\pm(t) a(\tilde{g}_\pm(t))^{-1}, \quad b_2^\pm(t) = \tilde{g}_\pm(t) \exp(t\eta_\pm)(\tilde{g}_\pm(t))^{-1},\\
&c^\pm(t) = \epsilon_\pm \tilde{g}_\pm(t) \exp(t\xi_\pm)(\tilde{g}_\pm(t))^{-1}
\end{aligned}$$

Then

$$\begin{aligned}
[a_1^\pm(t), b_1^\pm(t)] &= \epsilon_\pm \tilde{g}_\pm(t)[a, \exp(t\eta_\pm)](\epsilon_\pm \tilde{g}_\pm(1))^{-1}\\
&= \epsilon_\pm \tilde{g}_\pm(t) \exp(t\xi_\pm)(\epsilon_\pm \tilde{g}_\pm(t))^{-1} = c(t)(\epsilon_\pm^{-1}) = c(t)(\pm\epsilon_\pm),\\
[a_2^\pm(t), b_2^\pm(t)] &= \tilde{g}_\pm(t)[a, \exp(t\eta_\pm)](\tilde{g}_\pm(1))^{-1} = \tilde{g}_\pm(t) \exp(t\xi_\pm)(\tilde{g}_\pm(t))^{-1} = \epsilon_\pm^{-1} c(t),
\end{aligned}$$

so the image of $\gamma_\pm$ lies in $\tilde{V}^{\ell,1,\pm}_{Pin(n)_-}$. We have

$$\begin{aligned}
\gamma_\pm(0) &= (\epsilon_\pm a \epsilon_\pm^{-1}, 1, a, 1, 1, \ldots, 1, \epsilon_\pm) \in \pi_\pm^{-1}(\epsilon_\pm)\\
\gamma_\pm(1) &= (\epsilon_\pm g_\pm a(\epsilon_\pm g_\pm)^{-1}, \epsilon_\pm g_\pm \exp(\eta_\pm)(\epsilon_\pm g_\pm)^{-1}, g_\pm a(g_\pm)^{-1}, g_\pm \exp(\eta_\pm)(g_\pm)^{-1},\\
&\quad 1, \ldots, 1, \tilde{c}) \in \pi_\pm^{-1}(\tilde{c}).
\end{aligned}$$

$i = 2$. Define $\pi_\pm : \tilde{V}^{\ell,2,\pm 1}_{Pin(n)_-} \to Spin(n) \times Pin(n)_-$ by $(\tilde{a}_1, \tilde{b}_1, \ldots, \tilde{a}_\ell, \tilde{b}_\ell, \tilde{d}, \tilde{c}) \mapsto (\tilde{d}, \tilde{c})$. Note that $Spin(n)$ is simply connected and $\tilde{c}\tilde{d}\tilde{c}^{-1}\tilde{d}, -\tilde{c}\tilde{d}\tilde{c}^{-1}\tilde{d} \in Spin(n)$, so $\pi_\pm^{-1}(\tilde{d}, \tilde{c})$ is nonempty and connected for any $(\tilde{d}, \tilde{c}) \in Spin(n) \times Pin(n)_-$. Let $\epsilon_+ = 1$, and let $\epsilon_- = e_2e_3$. Then $e_1\epsilon_\pm e_1^{-1}\epsilon_\pm = e_1^{-1}\epsilon_\pm e_1 \epsilon_\pm = \pm 1$. It suffices to show that for any $(\tilde{d}, \tilde{c}) \in Spin(n) \times Pin(n)_-$, there is a path $\gamma_\pm : [0,1] \to \tilde{V}^{\ell,2,\pm 1}_{Pin(n)_-}$ such that $\gamma_\pm(0) \in \pi_\pm^{-1}(\epsilon_\pm, e_1)$ and $\gamma(1) \in \pi_\pm^{-1}(\tilde{d}, \tilde{c})$.

Let T be the maximal torus of $Spin(n)$, and let $\mathfrak{t}$ be the Lie algebra of T. Given $\tilde{d} \in Spin(n)$ and $\tilde{c} \in Pin(n)_-$, there exist $g_+, g_-, g \in Spin(n)$ such that and $\xi, \xi_+, \xi_- \in \mathfrak{t}$ such that

$$\tilde{c} = e_1 g \exp(\xi) g^{-1}, \tilde{d} = \epsilon_\pm g_\pm \exp(\xi_\pm)(g_\pm)^{-1}.$$

Let ω be the coxeter element. There are $\eta, \eta_+, \eta_- \in \mathfrak{t}$ such that

$$\omega \cdot \eta - \eta = \xi, \quad \omega \cdot \eta_\pm - \eta_\pm = \xi_\pm.$$

Let $a \in N(T) \subset Spin(n)$ be the corresponding element which represents $\omega \in W = N(T)/T$. Then

$$\begin{aligned} a\exp(t\eta)a^{-1}\exp(-t\eta) &= \exp(\omega \cdot t\eta - t\eta) = \exp(t\xi), \\ a\exp(t\eta_\pm)a^{-1}\exp(-t\eta_\pm) &= \exp(\omega \cdot t\eta_\pm - t\eta_\pm) = \exp(t\xi_\pm). \end{aligned}$$

Now since $Spin(n)$ is connected, there are paths $\tilde{g}, \tilde{g}_+, \tilde{g}_- : [0,1] \to Spin(n)$ such that

$$\tilde{g}(0) = \tilde{g}_+(0) = \tilde{g}_-(0) = 1, \quad \tilde{g}(1) = g, \quad \tilde{g}_\pm(1) = g_\pm.$$

Now define $\gamma : [0,1] \to Spin(n)^{2\ell+1} \times Pin(n)_-$ by

$$\gamma_\pm(t) = (a_1(t), b_1(t), a_2^\pm(t), b_2^\pm(t), a_3^\pm(t), b_3^\pm(t), a_4^\pm(t), b_4^\pm(t), 1, \ldots, 1, d^\pm(t), c(t))$$

where

$$\begin{aligned} &a_1(t) = e_1\tilde{g}(t)a(e_1\tilde{g}(t))^{-1}, \quad b_1(t) = e_1\tilde{g}(t)\exp(t\eta)(e_1\tilde{g}(t))^{-1}, \\ &a_2^\pm(t) = e_1\epsilon_\pm\tilde{g}_\pm(t)a(e_1\epsilon_\pm\tilde{g}_\pm(t))^{-1}, \quad b_2^\pm(t) = e_1\epsilon_\pm\tilde{g}_\pm(t)\exp(t\eta_\pm)(e_1\epsilon_\pm\tilde{g}_\pm(t))^{-1}, \\ &a_3^\pm(t) = e_1\epsilon_\pm\tilde{g}(t)a(e_1\epsilon_\pm\tilde{g}(t))^{-1}, \quad b_3^\pm(t) = e_1\epsilon_\pm\tilde{g}(t)\exp(-t\eta)(e_1\epsilon_\pm\tilde{g}(t))^{-1}, \\ &a_4^\pm(t) = \tilde{g}_\pm(t)a\tilde{g}_\pm(t)^{-1}, \quad b_4^\pm(t) = \tilde{g}_\pm(t)\exp(t\eta_\pm)\tilde{g}_\pm(t)^{-1}, \\ &c(t) = e_1\tilde{g}(t)\exp(t\xi)\tilde{g}(t)^{-1}, \quad d^\pm(t) = \epsilon_\pm\tilde{g}_\pm(t)\exp(t\xi_\pm)\tilde{g}_\pm(t)^{-1}. \end{aligned}$$

Then

$$\begin{aligned} &[a_1(t), b_1(t)] = e_1\tilde{g}(t)[a, \exp(t\eta)](e_1\tilde{g}(t))^{-1} = c(t)e_1^{-1}, \\ &[a_2^\pm(t), b_2^\pm(t)] = e_1\epsilon_\pm\tilde{g}_\pm(t)[a, \exp(t\eta_\pm)](e_1\epsilon_\pm\tilde{g}_\pm(t))^{-1} = e_1 d(t)(e_1\epsilon_\pm)^{-1} \\ &[a_3^\pm(t), b_3^\pm(t)] = e_1\epsilon_\pm\tilde{g}(t)[a, \exp(-t\eta)]\tilde{g}(t)^{-1}(e_1\epsilon_\pm)^{-1} = e_1\epsilon_\pm c(t)^{-1}(\pm\epsilon_\pm), \\ &[a_4^\pm(t), b_4^\pm(t)] = \tilde{g}_\pm(t)[a, \exp(t\eta_\pm)](\tilde{g}_\pm(t))^{-1} = \epsilon_\pm^{-1}d(t), \end{aligned}$$

so the image of $\gamma_\pm$ lies in $\tilde{V}^{\ell,2,\pm}_{Pin(n)_-}$. We have

$$\begin{aligned} \gamma_\pm(0) &= (e_1 a e_1^{-1}, 1, \ e_1\epsilon_\pm a(e_1\epsilon_\pm)^{-1}, 1, \ e_1\epsilon_\pm a(e_1\epsilon_\pm)^{-1}, 1, a, 1, 1, \ldots, 1, \epsilon_\pm, e_1) \\ &\quad \in \pi_\pm^{-1}(\epsilon_\pm, e_1) \\ \gamma_\pm(1) &= (e_1 g a(e_1 g)^{-1}, e_1 g\exp(\eta)(e_1 g)^{-1}, e_1\epsilon_\pm g_\pm a(e_1\epsilon_\pm g_\pm)^{-1}, \\ &\quad e_1\epsilon_\pm g_\pm\exp(\eta_\pm)(e_1\epsilon_\pm g_\pm)^{-1}, e_1\epsilon_\pm g a(e_1\epsilon_\pm g)^{-1}, e_1\epsilon_\pm g\exp(-\eta)(e_1\epsilon_\pm g)^{-1}, \\ &\quad g_\pm a g_\pm^{-1}, \ g_\pm\exp(\eta_\pm)g_\pm^{-1}, \ 1, \ldots, 1, \tilde{d}, \tilde{c}) \in \pi_\pm^{-1}(\tilde{d}, \tilde{c}) \end{aligned}$$

□

CHAPTER 5

Yang-Mills $SO(2n+1)$-Connections

The maximal torus of $SO(2n+1)$ consists of block diagonal matrices of the form

$$\mathrm{diag}(A_1, \ldots, A_n, I_1),$$

where $A_1, \ldots, A_n \in SO(2)$, and I_1 is the 1×1 identity matrix. The Lie algebra of the maximal torus consists of matrices of the form

$$2\pi \mathrm{diag}(t_1 J, \ldots, t_n J, 0I_1) = 2\pi \begin{pmatrix} 0 & -t_1 & & & & 0 & 0 \\ t_1 & 0 & & & & & 0 \\ & & \cdot & & & & \\ & & & \cdot & & & \\ & & & & 0 & -t_n & \\ 0 & & & & t_n & 0 & 0 \\ 0 & 0 & & & & 0 & 0 \end{pmatrix},$$

where

$$J = \begin{pmatrix} 0 & -1 \\ 1 & 0 \end{pmatrix}. \tag{5.1}$$

The fundamental Weyl chamber is

$$\overline{C}_0 = \{\sqrt{-1}\mathrm{diag}(t_1 J, \ldots, t_n J, 0I_1) \mid t_1 \geq t_2 \geq \cdots \geq t_n \geq 0\}.$$

In this chapter, we assume

$$n_1, \ldots, n_r \in \mathbb{Z}_{>0}, \quad n_1 + \cdots + n_r = n.$$

5.1. $SO(2n+1)$-connections on orientable surfaces

Let J_m denote the $2m \times 2m$ matrix $\mathrm{diag}(\underbrace{J, \ldots, J}_{m})$. Any $\mu \in \overline{C}_0$ is of the form

$$\mu = \sqrt{-1}\mathrm{diag}(\lambda_1 J_{n_1}, \ldots, \lambda_r J_{n_r}, 0I_1),$$

where $\lambda_1 > \cdots > \lambda_r \geq 0$.

Let $X_\mu = -2\pi\sqrt{-1}\mu$. Then

$$SO(2n+1)_{X_\mu} \cong \begin{cases} \Phi(U(n_1)) \times \cdots \times \Phi(U(n_r)), & \lambda_r > 0, \\ \Phi(U(n_1)) \times \cdots \times \Phi(U(n_{r-1})) \times SO(2n_r+1), & \lambda_r = 0, \end{cases}$$

where $\Phi : U(m) \hookrightarrow SO(2m)$ is the standard embedding defined as follows. Consider the $\mathbb{R}$-linear map $L : \mathbb{R}^{2m} \to \mathbb{C}^m$ given by

$$\begin{pmatrix} x_1 \\ y_1 \\ \vdots \\ x_m \\ y_m \end{pmatrix} \mapsto \begin{pmatrix} x_1 + \sqrt{-1}y_1 \\ \vdots \\ x_m + \sqrt{-1}y_m \end{pmatrix}.$$

We have $L^{-1} \circ (\sqrt{-1}I_m) \circ L(v) = J_m v$ for $v \in \mathbb{R}^{2m}$. If A is a $m \times m$ matrix, let $\Phi(A)$ be the $2m \times 2m$ matrix defined by

$$L^{-1} \circ A \circ L(v) = \Phi(A)(v), \quad v \in \mathbb{R}^{2m}. \tag{5.2}$$

Note that $A(\sqrt{-1}I_m) = (\sqrt{-1}I_m)A \Rightarrow J_m\Phi(A) = \Phi(A)J_m$.

Suppose that $(a_1, b_1, \ldots, a_\ell, b_\ell, X_\mu) \in X^{\ell,0}_{\mathrm{YM}}(SO(2n+1))$. Then

$$\exp(X_\mu) = \prod_{i=1}^{\ell} [a_i, b_i]$$

where $a_i,\ b_i \in SO(2n+1)_{X_\mu}$. This implies that $\exp(X_\mu) \in (SO(2n+1)_{X_\mu})_{ss}$, the semisimple part of $SO(2n+1)_{X_\mu}$:

$$(SO(2n+1)_{X_\mu})_{ss} = \begin{cases} \Phi(SU(n_1)) \times \cdots \times \Phi(SU(n_r)), & \lambda_r > 0, \\ \Phi(SU(n_1)) \times \cdots \times \Phi(SU(n_{r-1})) \times SO(2n_r+1), & \lambda_r = 0. \end{cases}$$

Thus

$$\begin{aligned} X_\mu &= 2\pi \mathrm{diag}\Big(\frac{k_1}{n_1} J_{n_1}, \ldots, \frac{k_r}{n_r} J_{n_r}, 0I_1\Big), \\ \mu &= \sqrt{-1}\mathrm{diag}\Big(\frac{k_1}{n_1} J_{n_1}, \ldots, \frac{k_r}{n_r} J_{n_r}, 0I_1\Big), \end{aligned}$$

where

$$k_1, \ldots, k_r \in \mathbb{Z}, \quad \frac{k_1}{n_1} > \cdots > \frac{k_r}{n_r} \geq 0.$$

This agrees with Section 3.4.2.

Recall that for each μ, the representation variety is

$$V^{\ell,0}_{\mathrm{YM}}(SO(2n+1))_\mu = \{(a_1, b_1, \ldots, a_\ell, b_\ell) \in (SO(2n+1)_{X_\mu})^{2\ell} \mid \prod_{i=1}^{\ell}[a_i, b_i] = \exp(X_\mu)\}.$$

For $i = 1, \cdots, \ell$, write

$$\begin{aligned} a_i &= \mathrm{diag}(A^i_1, \ldots, A^i_r, I_1),\ b_i = \mathrm{diag}(B^i_1, \ldots, B^i_r, I_1), && \text{when } k_r > 0, \\ a_i &= \mathrm{diag}(A^i_1, \ldots, A^i_r),\ b_i = \mathrm{diag}(B^i_1, \ldots, B^i_r), && \text{when } k_r = 0, \end{aligned}$$

where $A^i_j,\ B^i_j \in \Phi(U(n_j))$ for $j = 1, \ldots, r-1$, and

$$A^i_r,\ B^i_r \in \begin{cases} \Phi(U(n_r)), & \text{when } k_r > 0, \\ SO(2n_r+1), & \text{when } k_r = 0. \end{cases}$$

Let

$$\hat{J}_t = \exp(2\pi t J) = \begin{pmatrix} \cos(2\pi t) & -\sin(2\pi t) \\ \sin(2\pi t) & \cos(2\pi t) \end{pmatrix},$$

and let

$$T_{n,k} = \Phi(e^{2\pi\sqrt{-1}k/n} I_n) = \mathrm{diag}(\underbrace{\hat{J}_{k/n}, \ldots, \hat{J}_{k/n}}_{n}) \in SO(2n). \tag{5.3}$$

For $j = 1, \ldots, r-1$, define

$$\begin{aligned} V_j &= \Big\{(A_j^1, B_j^1, \ldots, A_j^\ell, B_j^\ell) \in \Phi(U(n_j))^{2\ell} \mid \prod_{i=1}^{\ell}[A_j^i, B_j^i] = T_{n_j,k_j}\Big\} \\ &\overset{\Phi}{\cong} \Big\{(A_j^1, B_j^1, \ldots, A_j^\ell, B_j^\ell) \in U(n_j)^{2\ell} \mid \prod_{i=1}^{\ell}[A_j^i, B_j^i] = e^{2\pi\sqrt{-1}k_j/n_j} I_{n_j})\Big\} \\ &\cong X_{\mathrm{YM}}^{\ell,0}(U(n_j))_{-\frac{k_j}{n_j}, \ldots, -\frac{k_j}{n_j}}. \end{aligned} \tag{5.4}$$

If $k_r > 0$, define V_r by (5.4). If $k_r = 0$, define

$$\begin{aligned} V_r &= \{(A_r^1, B_r^1, \ldots, A_r^\ell, B_r^\ell) \in SO(2n_r+1)^{2\ell} \mid \prod_{i=1}^{\ell}[A_r^i, B_r^i] = I_{2n_r+1}\Big\} \\ &\cong X_{\mathrm{flat}}^{\ell,0}(SO(2n_r+1)). \end{aligned}$$

Then $V_{\mathrm{YM}}^{\ell,0}(SO(2n+1))_\mu = \prod_{j=1}^{r} V_j$. We have a homeomorphism

$$V_{\mathrm{YM}}^{\ell,0}(SO(2n+1))_\mu / SO(2n+1)_{X_\mu} = \begin{cases} \displaystyle\prod_{j=1}^{r}(V_j/U(n_j)), & k_r > 0, \\ \displaystyle\prod_{j=1}^{r-1}(V_j/U(n_j)) \times V_r/SO(2n_r+1), & k_r = 0, \end{cases}$$

and a homotopy equivalence

$$V_{\mathrm{YM}}^{\ell,0}(SO(2n+1))_\mu{}^{hSO(2n+1)_{X_\mu}} \sim \begin{cases} \prod_{j=1}^{r} V_j{}^{hU(n_j)}, & k_r > 0, \\ \prod_{j=1}^{r-1} V_j{}^{hU(n_j)} \times V_r{}^{hSO(2n_r+1)}, & k_r = 0. \end{cases}$$

NOTATION 5.1. Suppose that $m \geq 3$. Let Σ be a closed, orientable or nonorientable surface. Let $P_{SO(m)}^{+1}$ and $P_{SO(m)}^{-1}$ denote the principal $SO(m)$-bundle on Σ with $w_2(P_{SO(m)}^{+1}) = 0$ and $w_2(P_{SO(m)}^{-1}) = 1$ respectively in $H^2(\Sigma; \mathbb{Z}/2\mathbb{Z}) \cong \mathbb{Z}/2\mathbb{Z}$. Let $\mathcal{N}(\Sigma)_{SO(m)}^{\pm 1}$ denote the space of Yang-Mills connections on $P_{SO(m)}^{\pm 1}$, and let $\mathcal{N}_0(\Sigma)_{SO(m)}^{\pm}$ denote the space of flat connections on $P_{SO(m)}^{\pm 1}$.

For $i = 0, 1, 2$, we have

$$X_{\mathrm{YM}}^{\ell,i}(SO(m)) = X_{\mathrm{YM}}^{\ell,i}(SO(m))^{+1} \cup X_{\mathrm{YM}}^{\ell,i}(SO(m))^{-1}$$

where

$$X_{\mathrm{YM}}^{\ell,i}(SO(m))^{\pm 1} \cong \mathcal{N}(\Sigma_i^\ell)_{SO(m)}^{\pm 1} / \mathcal{G}_0(P_{SO(m)}^{\pm 1}),$$

and

$$X_{\mathrm{flat}}^{\ell,i}(SO(m)) = X_{\mathrm{flat}}^{\ell,i}(SO(m))^{+1} \cup X_{\mathrm{flat}}^{\ell,i}(SO(m))^{-1}$$

where

$$X_{\mathrm{flat}}^{\ell,i}(SO(m))^{\pm 1} = \mathcal{N}_0(\Sigma_i^\ell)_{SO(m)}^{\pm 1} / \mathcal{G}_0(P_{SO(m)}^{\pm 1})$$

is nonempty and connected for $\ell \geq 1$. Let

$$X_{\mathrm{YM}}^{\ell,i}(SO(m))_\mu^{\pm 1} = X_{\mathrm{YM}}^{\ell,i}(SO(m))_\mu \cap X_{\mathrm{YM}}^{\ell,i}(SO(m))^{\pm 1}$$

be the representation varieties for Yang-Mills connections of type μ on $P^{\pm 1}_{SO(m)}$. Let

$$\mathcal{M}(\Sigma, P^{\pm 1}_{SO(m)}) = X^{\ell,i}_{\mathrm{flat}}(SO(m))^{\pm 1}/SO(m)$$

be the moduli space of gauge equivalence classes of flat connections on $P^{\pm 1}_{SO(m)}$ over Σ. Let

$$\mathcal{M}(\Sigma^\ell_0, P^{n,k}) = X^{\ell,0}_{\mathrm{YM}}(U(n))_{\frac{k}{n},\dots,\frac{k}{n}}/U(n)$$

be the moduli space of gauge equivalence classes of central Yang-Mills connections on a degree k principal $U(n)$-bundle over Σ^ℓ_0. Recall that there is no flat connection on a degree $k \neq 0$ principal $U(n)$-bundle over Σ^ℓ_0.

We have seen that $V^{\ell,0}_{\mathrm{YM}}(SO(2n+1))_\mu = \prod_{j=1}^r V_j$ is connected for $k_r > 0$ and disconnected with two connected components for $k_r = 0$. To determine the underlying topological type of the $SO(2n+1)$-bundle, let us consider the group homomorphism

$$\phi_\mu : \pi_1(SO(2n+1)_{X_\mu}) \to \pi_1(SO(2n+1)) \cong \mathbb{Z}/2\mathbb{Z}$$

induced by the inclusion $SO(2n+1)_{X_\mu} \hookrightarrow SO(2n+1)$. We have

$$\pi_1(SO(2n{+}1)_{X_\mu}) \cong \begin{cases} \prod_{j=1}^{r} \pi_1(U(n_j)) \cong \mathbb{Z}^r, & \lambda_r > 0,\\ \prod_{j=1}^{r-1} \pi_1(U(n_j)) \times \pi_1(SO(2n_r+1)) \cong \mathbb{Z}^{r-1} \times \mathbb{Z}/2\mathbb{Z}, & \lambda_r = 0, \end{cases}$$

and

$$\phi_\mu(k_1,\dots,k_r) = k_1 + \cdots + k_r \pmod 2.$$

Thus, for $k_r > 0$, $V^{\ell,0}_{\mathrm{YM}}(SO(2n+1))_\mu$ is from the trivial $SO(2n+1)$-bundle if and only if $k_1 + \cdots + k_r = 0 \pmod 2$; and for $k_r = 0$, $V^{\ell,0}_{\mathrm{YM}}(SO(2n+1))_\mu$ has two connected components $V^{\ell,0}_{\mathrm{YM}}(SO(2n+1))^+_\mu$ and $V^{\ell,0}_{\mathrm{YM}}(SO(2n+1))^-_\mu$, where

$$\begin{aligned} V^{\ell,0}_{\mathrm{YM}}(SO(2n+1))^+_\mu &= \prod_{j=1}^{r-1} V_j \times X^{\ell,0}_{\mathrm{flat}}(SO(2n_r+1))^{(-1)^{k_1+\cdots+k_{r-1}}},\\ V^{\ell,0}_{\mathrm{YM}}(SO(2n+1))^-_\mu &= \prod_{j=1}^{r-1} V_j \times X^{\ell,0}_{\mathrm{flat}}(SO(2n_r+1))^{(-1)^{k_1+\cdots+k_{r-1}+1}}. \end{aligned}$$

To simplify the notation, we write

$$\mu = (\mu_1,\dots,\mu_n) = \Big(\underbrace{\frac{k_1}{n_1},\dots,\frac{k_1}{n_1}}_{n_1},\dots,\underbrace{\frac{k_r}{n_r},\dots,\frac{k_r}{n_r}}_{n_r}\Big)$$

instead of

$$\sqrt{-1}\mathrm{diag}\Big(\frac{k_1}{n_1}J_{n_1},\dots,\frac{k_r}{n_r}J_{n_r}, 0I_1\Big).$$

Let

$$I_{SO(2n+1)} = \Big\{\mu = \Big(\underbrace{\frac{k_1}{n_1},\dots,\frac{k_1}{n_1}}_{n_1},\dots,\underbrace{\frac{k_r}{n_r},\dots,\frac{k_r}{n_r}}_{n_r}\Big) \,\Big|\, \begin{matrix} n_j \in \mathbb{Z}_{>0},\ n_1+\cdots+n_r = n\\ k_j \in \mathbb{Z},\ \frac{k_1}{n_1} > \cdots > \frac{k_r}{n_r} \geq 0 \end{matrix} \Big\},$$

$$\begin{aligned} I^{\pm 1}_{SO(2n+1)} &= \{\mu \in I_{SO(2n+1)} \mid \mu_n > 0, (-1)^{k_1+\cdots+k_r} = \pm 1\}, \\ I^{0}_{SO(2n+1)} &= \{\mu \in I_{SO(2n+1)} \mid \mu_n = 0\}. \end{aligned}$$

From the discussion above, we conclude:

PROPOSITION 5.2. *Suppose that $\ell \geq 1$. Let*

$$\mu = \Big(\underbrace{\frac{k_1}{n_1}, \ldots, \frac{k_1}{n_1}}_{n_1}, \ldots, \underbrace{\frac{k_r}{n_r}, \ldots, \frac{k_r}{n_r}}_{n_r}\Big) \in I_{SO(2n+1)}. \tag{5.5}$$

(i) *If $\mu \in I^{\pm 1}_{SO(2n+1)}$, then $X^{\ell,0}_{\mathrm{YM}}(SO(2n+1))_\mu = X^{\ell,0}_{\mathrm{YM}}(SO(2n+1))^{\pm 1}_\mu$ is nonempty and connected. We have a homeomorphism*

$$X^{\ell,0}_{\mathrm{YM}}(SO(2n+1))_\mu / SO(2n+1) \cong \prod_{j=1}^{r} \mathcal{M}(\Sigma^\ell_0, P^{n_j,-k_j})$$

and a homotopy equivalence

$$X^{\ell,0}_{\mathrm{YM}}(SO(2n+1))_\mu^{\,hSO(2n+1)} \sim \prod_{j=1}^{r} \Big(X^{\ell,0}_{\mathrm{YM}}(U(n_j))_{-\frac{k_j}{n_j},\ldots,-\frac{k_j}{n_j}} \Big)^{hU(n_j)}.$$

(ii) *If $\mu \in I^{0}_{SO(2n+1)}$, then $X^{\ell,0}_{\mathrm{YM}}(SO(2n+1))_\mu$ has two connected components (from both bundles over Σ^ℓ_0)*

$$X^{\ell,0}_{\mathrm{YM}}(SO(2n+1))^{+1}_\mu \quad \text{and} \quad X^{\ell,0}_{\mathrm{YM}}(SO(2n+1))^{-1}_\mu.$$

We have a homeomorphism

$$X^{\ell,0}_{\mathrm{YM}}(SO(2n+1))^{\pm 1}_\mu / SO(2n+1) \cong \prod_{j=1}^{r-1} \mathcal{M}(\Sigma^\ell_0, P^{n_j,-k_j}) \times \mathcal{M}\Big(\Sigma^\ell_0, P^{\pm(-1)^{k_1+\cdots+k_{r-1}}}_{SO(2n_r+1)}\Big)$$

and a homotopy equivalence

$$\Big(X^{\ell,0}_{\mathrm{YM}}(SO(2n+1))^{\pm}_\mu\Big)^{hSO(2n+1)} \sim \prod_{j=1}^{r-1} \Big(X^{\ell,0}_{\mathrm{YM}}(U(n_j))_{-\frac{k_j}{n_j},\ldots,-\frac{k_j}{n_j}} \Big)^{hU(n_j)} \times \Big(X^{\ell,0}_{\mathrm{flat}}(SO(2n_r+1))^{\pm(-1)^{k_1+\cdots+k_{r-1}}}\Big)^{hSO(2n_r+1)}.$$

PROPOSITION 5.3. *Suppose that $\ell \geq 1$. The connected components of the representation variety $X^{\ell,0}_{\mathrm{YM}}(SO(2n+1))^{\pm 1}$ are*

$$\{X^{\ell,0}_{\mathrm{YM}}(SO(2n+1))_\mu \mid \mu \in I^{\pm 1}_{SO(2n+1)}\} \cup \{X^{\ell,0}_{\mathrm{YM}}(SO(2n+1))^{\pm 1}_\mu \mid \mu \in I^{0}_{SO(2n+1)}\}.$$

The following is an immediate consequence of Proposition 5.2.

THEOREM 5.4. *Suppose that $\ell \geq 1$, and let μ be as in (5.5).*

(i) *If $\mu \in I^{\pm 1}_{SO(2n+1)}$, then*

$$P^{SO(2n+1)}_t \Big(X^{\ell,0}_{\mathrm{YM}}(SO(2n+1))_\mu\Big) = \prod_{j=1}^{r} P^{U(n_j)}_t \Big(X^{\ell,0}_{\mathrm{YM}}(U(n_i))_{-\frac{k_j}{n_j},\ldots,-\frac{k_j}{n_j}} \Big).$$

(ii) *If* $\mu \in I^0_{SO(2n+1)}$, *then*

$$P_t^{SO(2n+1)}\left(X_{\mathrm{YM}}^{\ell,0}(SO(2n+1))^{\pm 1}_{\mu}\right) = \prod_{j=1}^{r-1} P_t^{U(n_j)}\left(X_{\mathrm{YM}}^{\ell,0}(U(n_j))_{-\frac{k_j}{n_j},\dots,-\frac{k_j}{n_j}}\right) \times$$
$$P_t^{SO(2n_r+1)}\left(X_{\mathrm{flat}}^{\ell,0}(SO(2n_r+1))^{\pm(-1)^{k_1+\cdots+k_{r-1}}}\right).$$

5.2. Equivariant Poincaré series

Recall from Chapter 3.4.2:

$$\Delta = \{\alpha_i = \theta_i - \theta_{i+1} \mid i = 1,\dots,n-1\} \cup \{\alpha_n = \theta_n\}$$
$$\Delta^\vee = \{\alpha_i^\vee = e_i - e_{i+1} \mid i = 1,\dots,n-1\} \cup \{\alpha_n^\vee = 2e_n\}$$
$$\pi_1(H) = \bigoplus_{i=1}^{n} \mathbb{Z}e_i, \quad \Lambda = \bigoplus_{i=1}^{n-1} \mathbb{Z}(e_i - e_{i+1}) \oplus \mathbb{Z}(2e_n),$$
$$\pi_1(SO(2n+1)) = \langle e_n \rangle \cong \mathbb{Z}/2\mathbb{Z}$$

We now apply Theorem 4.4 to the case $G_{\mathbb{R}} = SO(2n+1)$.

$$\varpi_{\alpha_i} = \theta_1 + \cdots + \theta_i, \quad i = 1,\dots,n-1, \quad \varpi_{\alpha_n} = \frac{1}{2}(\theta_1 + \cdots + \theta_n)$$
$$\varpi_{\alpha_i}(ke_n) = \begin{cases} 0 & i < n \\ k/2 & i = n \end{cases}$$

Case 1. $\alpha_n \in I$:

$$I = \{\alpha_{n_1}, \alpha_{n_1+n_2}, \dots, \alpha_{n_1+\cdots+n_{r-1}}, \alpha_n\}$$
$$L^I = GL(n_1,\mathbb{C}) \times \cdots \times GL(n_r,\mathbb{C}), \quad n_1 + \cdots + n_r = n$$
$$\dim_{\mathbb{C}} \mathfrak{z}_{L^I} - \dim_{\mathbb{C}} \mathfrak{z}_{SO(2n+1,\mathbb{C})} = r, \quad \dim_{\mathbb{C}} U^I = \sum_{1\le i<j\le r} n_i n_j + \frac{n(n+1)}{2},$$
$$\rho^I = \frac{1}{2}\sum_{i=1}^{r}\left(n - 2\sum_{j=1}^{i} n_j + n_i\right)\left(\sum_{j=1}^{n_i} \theta_{n_1+\cdots+n_{i-1}+j}\right) + \frac{n}{2}(\theta_1 + \cdots + \theta_n)$$
$$\langle \rho^I, \alpha^\vee_{n_1+\cdots+n_i}\rangle = \frac{n_i + n_{i+1}}{2} \text{ for } i = 1,\dots,r-1, \quad \langle \rho^I, \alpha_n^\vee \rangle = n_r$$

Case 2. $\alpha_n \notin I$:

$$I = \{\alpha_{n_1}, \alpha_{n_1+n_2}, \dots, \alpha_{n_1+\cdots+n_{r-1}}\}$$
$$L^I = GL(n_1,\mathbb{C}) \times \cdots \times GL(n_{r-1},\mathbb{C}) \times SO(2n_r+1,\mathbb{C}), \quad n_1 + \cdots + n_r = n$$
$$\dim_{\mathbb{C}} \mathfrak{z}_{L^I} - \dim_{\mathbb{C}} \mathfrak{z}_{SO(2n+1,\mathbb{C})} = r - 1,$$
$$\dim_{\mathbb{C}} U^I = \sum_{1\le i<j\le r} n_i n_j + \frac{n(n+1) - n_r(n_r+1)}{2},$$
$$\rho^I = \frac{1}{2}\sum_{i=1}^{r}\left(n - 2\sum_{j=1}^{i} n_j + n_i\right)\left(\sum_{j=1}^{n_i} \theta_{n_1+\cdots+n_{i-1}+j}\right)$$
$$+\frac{n}{2}(\theta_1 + \cdots + \theta_{n_1+\cdots+n_{r-1}}) + \frac{n-n_r}{2}(\theta_{n_1+\cdots+n_{r-1}+1} + \cdots + \theta_n)$$
$$\langle \rho^I, \alpha^\vee_{n_1+\cdots+n_i}\rangle = \frac{n_i + n_{i+1}}{2} \text{ for } i = 1,\dots,r-2, \quad \langle \rho^I, \alpha^\vee_{n_1+\cdots+n_{r-1}}\rangle = \frac{n_{r-1}}{2} + n_r$$

We have the following closed formula for the $SO(2n+1)$-equivariant poincaré series of the representation of flat $SO(2n+1)$-connections:

THEOREM 5.5.

$$
\begin{aligned}
&P_t^{SO(2n+1)}(X_{\text{flat}}^{\ell,0}(SO(2n+1))^{(-1)^k}) \\
&= \sum_{r=1}^{n} \sum_{\substack{n_1,\dots,n_r \in \mathbb{Z}_{>0} \\ \sum n_j = n}} \Bigg((-1)^r \prod_{i=1}^{r} \frac{\prod_{j=1}^{n_i}(1+t^{2j-1})^{2\ell}}{(1-t^{2n_i})\prod_{j=1}^{n_i-1}(1-t^{2j})^2} \\
&\quad \cdot \frac{t^{(\ell-1)(2\sum_{i<j} n_i n_j + n(n+1))}}{\left[\prod_{i=1}^{r-1}(1-t^{2(n_i+n_{i+1})})\right](1-t^{4n_r})} \cdot t^{2\sum_{i=1}^{r-1}(n_i+n_{i+1})+4n_r\langle k/2\rangle} \\
&\quad +(-1)^{r-1} \prod_{i=1}^{r-1} \frac{\prod_{j=1}^{n_i}(1+t^{2j-1})^{2\ell}}{(1-t^{2n_i})\prod_{j=1}^{n_i-1}(1-t^{2j})^2} \cdot \frac{\prod_{j=1}^{n_r}(1+t^{4j-1})^{2\ell}}{\prod_{j=1}^{2n_r}(1-t^{2j})} \\
&\quad \cdot \frac{t^{(\ell-1)(2\sum_{i<j} n_i n_j + n(n+1) - n_r(n_r+1))}}{\left[\prod_{i=1}^{r-2}(1-t^{2(n_i+n_{i+1})})\right](1-\epsilon(r)t^{2n_{r-1}+4n_r})} t^{2\sum_{i=1}^{r-1}(n_i+n_{i+1})+2\epsilon(r)n_r} \Bigg)
\end{aligned}
$$

where.

$$
\epsilon(r) = \begin{cases} 0 & r = 1 \\ 1 & r > 1 \end{cases}
$$

REMARK 5.6. We have

$$
P_t^{SO(2n+1)}(X_{\text{flat}}^{\ell,0}(SO(2n+1))^{+1}) = P_t^{Spin(2n+1)}(X_{\text{flat}}^{\ell,0}(Spin(2n+1))),
$$

so Theorem 5.5 also gives a formula for $X_{\text{flat}}^{\ell,0}(Spin(2n+1))$.

EXAMPLE 5.7.

$$
\begin{aligned}
&P_t^{SO(3)}(X_{\text{flat}}^{\ell,0}(SO(3))^{+1}) = P_t^{Spin(3)}(X_{\text{flat}}^{\ell,0}(Spin(3))) \\
&= -\frac{(1+t)^{2\ell}t^{2\ell+2}}{(1-t^2)(1-t^4)} + \frac{(1+t^3)^{2\ell}}{(1-t^2)(1-t^4)} \\
&P_t^{SO(3)}(X_{\text{flat}}^{\ell,0}(SO(3))^{-1}) \\
&= -\frac{(1+t)^{2\ell}t^{2\ell}}{(1-t^2)(1-t^4)} + \frac{(1+t^3)^{2\ell}}{(1-t^2)(1-t^4)}
\end{aligned}
$$

Note that $Spin(3) = SU(2)$, so

$$
P_t^{Spin(3)}(X_{\text{flat}}^{\ell,0}(Spin(3))) = P_t^{SU(2)}(X_{\text{flat}}^{\ell,0}(SU(2)))
$$

as expected, where $P_t^{SU(2)}(X_{\text{flat}}^{\ell,0}(SU(2))$ is calculated in Example 4.7.

EXAMPLE 5.8.

$$
\begin{aligned}
&P_t^{SO(5)}(X_{\text{flat}}^{\ell,0}(SO(5))^{+1}) = P_t^{Spin(5)}(X_{\text{flat}}^{\ell,0}(Spin(5)))\\
= & -\frac{(1+t)^{2\ell}(1+t^3)^{2\ell}t^{6\ell+2}}{(1-t^2)^2(1-t^4)(1-t^8)} + \frac{(1+t^3)^{2\ell}(1+t^7)^{2\ell}}{(1-t^2)(1-t^4)(1-t^6)(1-t^8)}\\
&+\frac{(1+t)^{4\ell}t^{8\ell}}{(1-t^2)^2(1-t^4)^2} - \frac{(1+t)^{2\ell}(1+t^3)^{2\ell}t^{6\ell}}{(1-t^2)^2(1-t^4)(1-t^6)}\\
&P_t^{SO(5)}(X_{\text{flat}}^{\ell,0}(SO(5))^{-1})\\
= & -\frac{(1+t)^{2\ell}(1+t^3)^{2\ell}t^{6\ell-2}}{(1-t^2)^2(1-t^4)(1-t^8)} + \frac{(1+t^3)^{2\ell}(1+t^7)^{2\ell}}{(1-t^2)(1-t^4)(1-t^6)(1-t^8)}\\
&+\frac{(1+t)^{4\ell}t^{8\ell-2}}{(1-t^2)^2(1-t^4)^2} - \frac{(1+t)^{2\ell}(1+t^3)^{2\ell}t^{6\ell}}{(1-t^2)^2(1-t^4)(1-t^6)}
\end{aligned}
$$

5.3. $SO(2n+1)$-connections on nonorientable surfaces

We have $\overline{C}_0^\tau = \overline{C}_0$ (any $\mu \in \overline{C}_0$ is conjugate to $-\mu$). Any $\mu \in \overline{C}_0^\tau$ is of the form

$$\mu = \sqrt{-1}\mathrm{diag}(\lambda_1 J_{n_1}, \ldots, \lambda_r J_{n_r}, 0I_1)$$

where $\lambda_1 > \cdots > \lambda_r \geq 0$. We have

$$SO(2n+1)_{X_\mu} \cong \begin{cases} \Phi(U(n_1)) \times \cdots \times \Phi(U(n_r)), & \lambda_r > 0, \\ \Phi(U(n_1)) \times \cdots \times \Phi(U(n_{r-1})) \times SO(2n_r+1), & \lambda_r = 0, \end{cases}$$

where $X_\mu = -2\pi\sqrt{-1}\mu$, and $\Phi : U(m) \hookrightarrow SO(2m)$ is the standard embedding.

Given $\mu \in \overline{C}_0$, define

$$\epsilon_\mu = \begin{cases} \mathrm{diag}\,(H_n, (-1)^n I_1), & \lambda_r > 0, \\ \mathrm{diag}\,(H_{n-n_r}, (-1)^{n-n_r} I_1, I_{2n_r}), & \lambda_r = 0. \end{cases}$$

Then $\mathrm{Ad}(\epsilon_\mu)X_\mu = -X_\mu$. Suppose that

$$(a_1, b_1, \ldots, a_\ell, b_\ell, \epsilon_\mu c', X_\mu/2) \in X_{\mathrm{YM}}^{\ell,1}(SO(2n+1)).$$

Then

$$\exp(X_\mu/2)\epsilon_\mu c' \epsilon_\mu c' = \prod_{i=1}^{\ell}[a_i, b_i]$$

where

$$a_i,\ b_i,\ c' \in \begin{cases} \Phi(U(n_1)) \times \cdots \times \Phi(U(n_r)), & \lambda_r > 0, \\ \Phi(U(n_1)) \times \cdots \times \Phi(U(n_{r-1})) \times SO(2n_r+1), & \lambda_r = 0. \end{cases}$$

We first assume that $\lambda_r > 0$. Let $L : \mathbb{R}^{2n} \to \mathbb{C}^n$ defined as in Section 5.1. Define

$$
\begin{aligned}
X'_\mu &= L \circ 2\pi\mathrm{diag}(\lambda_1 J_{n_1}, \ldots, \lambda_r J_{n_r}) \circ L^{-1}\\
&= 2\pi\sqrt{-1}\mathrm{diag}(\lambda_1 I_{n_1}, \ldots, \lambda_r I_{n_r}) \in \mathfrak{u}(n_1) \times \cdots \times \mathfrak{u}(n_r).
\end{aligned}
$$

We have $L \circ H_n \circ L^{-1}(v) = \bar{v}$ for $v \in \mathbb{C}^n$, where $\bar{v}$ is the complex conjugate of v. So

$$
\begin{aligned}
L \circ H_n \Phi(c') H_n \circ L^{-1}(v) &= (L \circ H_n \circ L^{-1})(L \circ \Phi(c') \circ L^{-1})(L \circ H_n \circ L^{-1})(v)\\
&= (L \circ H_n \circ L^{-1})c'\bar{v} = \overline{c'\bar{v}} = \bar{c}'v.
\end{aligned}
$$

So the condition on X'_μ is

$$\exp(X'_\mu/2)\bar{c}'c' = \prod_{i=1}^{\ell}[a_i, b_i] \in SU(n_1) \times \cdots \times SU(n_r),$$

where $a_i,\ b_i,\ c' \in U(n_1) \times \cdots \times U(n_r)$, and $\bar{c}'$ is the complex conjugate of c'. In order that this is nonempty, we need $1 = \det(e^{\pi\sqrt{-1}\lambda_j} I_{n_j})$, or equivalently

$$\lambda_j = \frac{2k_j}{n_j} \quad k_j,\ n_j \in \mathbb{Z}_{>0} \tag{5.6}$$

for $j = 1, \ldots, r$.

When $\lambda_r = 0$, the above argument gives the condition (5.6) for $j = 1, \ldots, r-1$.

Similarly, suppose that

$$(a_1, b_1, \ldots, a_\ell, b_\ell, d, \epsilon_\mu c', X_\mu/2) \in X^{\ell,1}_{\mathrm{YM}}(SO(2n+1)).$$

Then

$$\exp(X_\mu/2)(\epsilon_\mu c')d(\epsilon_\mu c')^{-1}d = \prod_{i=1}^{\ell}[a_i, b_i],$$

where

$$a_i,\ b_i,\ d,\ c' \in \begin{cases} \Phi(U(n_1)) \times \cdots \times \Phi(U(n_r)), & \lambda_r > 0, \\ \Phi(U(n_1)) \times \cdots \times \Phi(U(n_{r-1})) \times SO(2n_r+1), & \lambda_r = 0. \end{cases}$$

When $\lambda_r > 0$, the condition on X'_μ is

$$\exp(X'_\mu/2)\bar{c}'\bar{d}\bar{c}'^{-1}d = \prod_{i=1}^{\ell}[a_i, b_i] \in SU(n_1) \times \cdots \times SU(n_r).$$

Again, we need $1 = \det(e^{\pi\sqrt{-1}\lambda_j} I_{n_j})$, or equivalently (5.6). When $\lambda_r = 0$ we get condition (5.6) for $j = 1, \ldots, r-1$.

We conclude that for nonorientable surfaces,

$$\mu = \sqrt{-1}\mathrm{diag}\Big(\frac{2k_1}{n_1}J_{n_1}, \ldots, \frac{2k_r}{n_r}J_{n_r}, 0I_1\Big), \text{ where } k_1, \ldots, k_r \in \mathbb{Z},\ \frac{k_1}{n_1} > \cdots > \frac{k_r}{n_r} \geq 0.$$

Recall that for *orientable* surfaces we have

$$\mu = \sqrt{-1}\mathrm{diag}\Big(\frac{k_1}{n_1}J_{n_1}, \ldots, \frac{k_r}{n_r}J_{n_r}, 0I_1\Big), \text{ where } k_1, \ldots, k_r \in \mathbb{Z},\ \frac{k_1}{n_1} > \cdots > \frac{k_r}{n_r} \geq 0.$$

For each μ, define ϵ_μ-reduced representation varieties

$$\begin{aligned} V^{\ell,1}_{\mathrm{YM}}(SO(2n+1))_\mu &= \{(a_1, b_1, \ldots, a_\ell, b_\ell, c') \in (SO(2n+1)_{X_\mu})^{2\ell+1} \mid \\ &\qquad \prod_{i=1}^{\ell}[a_i, b_i] = \exp(X_\mu/2)\epsilon_\mu c'\epsilon_\mu c'\}, \\ V^{\ell,2}_{\mathrm{YM}}(SO(2n+1))_\mu &= \{(a_1, b_1, \ldots, a_\ell, b_\ell, d, c') \in (SO(2n+1)_{X_\mu})^{2\ell+2} \mid \\ &\qquad \prod_{i=1}^{\ell}[a_i, b_i] = \exp(X_\mu/2)\epsilon_\mu c'd(\epsilon_\mu c')^{-1}d\}. \end{aligned}$$

For $i = 1, \ldots, \ell$, write

$$a_i = \mathrm{diag}(A^i_1, \ldots, A^i_r, I_1), \quad b_i = \mathrm{diag}(B^i_1, \ldots, B^i_r, I_1),$$
$$c' = \mathrm{diag}(C_1, \ldots, C_r, I_1), \quad d = \mathrm{diag}(D_1, \ldots, D_r, I_1),$$

when $k_r > 0$, and write

$$a_i = \mathrm{diag}(A_1^i, \dots, A_r^i), \quad b_i = \mathrm{diag}(B_1^i, \dots, B_r^i),$$
$$c' = \mathrm{diag}(C_1, \dots, C_r), \quad d = \mathrm{diag}(D_1, \dots, D_r),$$

when $k_r = 0$, where $A_j^i,\ B_j^i,\ D_j,\ C_j \in \Phi(U(n_j))$ for $j = 1, \dots, r-1$, and

$$A_r^i,\ B_r^i,\ D_r,\ C_r \in \begin{cases} \Phi(U(n_r)) & \text{when } k_r > 0, \\ SO(2n_r+1) & \text{when } k_r = 0. \end{cases}$$

$i = 1$. Let $T_{n,k}$ be defined as in (5.3), and let $\epsilon_j = \mathrm{diag}(H_{n_j})$. For $j = 1, \dots, r-1$, define

(5.7)

$$\begin{aligned} V_j &= \Big\{(A_j^1, B_j^1, \dots, A_j^\ell, B_j^\ell, C_j) \in \Phi(U(n_j))^{2\ell+1} \mid \prod_{i=1}^{\ell}[A_j^i, B_j^i] = T_{n_j,k_j}\epsilon_j C_j \epsilon_j C_j\Big\} \\ &\overset{\Phi}{\cong} \Big\{(A_j^1, B_j^1, \dots, A_j^\ell, B_j^\ell, C_j) \in U(n_j)^{2\ell+1} \mid \prod_{i=1}^{\ell}[A_j^i, B_j^i] = e^{2\pi\sqrt{-1}k_j/n_j}\bar{C}_j C_j\Big\} \\ &\cong \tilde{V}^{\ell,1}_{n_j,-k_j} \end{aligned}$$

where $\tilde{V}^{\ell,1}_{n_j,-k_j}$ is the twisted representation variety defined in (4.7) of Section 4.6. $\tilde{V}^{\ell,1}_{n_j,-k_j}$ is nonempty if $\ell \geq 1$. We have shown that $\tilde{V}^{\ell,1}_{n_j,-k_j}$ is connected if $\ell \geq 2$ (Proposition 4.13).

When $k_r > 0$, define V_r by (5.7). When $k_r = 0$, define

$$(5.8)\quad V_r = \Big\{(A_r^1, B_r^1, \dots, A_r^\ell, B_r^\ell, C_r) \in SO(2n_r+1)^{2\ell+1} \mid \prod_{i=1}^{\ell}[A_r^i, B_r^i] = (\epsilon C_r)^2\Big\},$$

where $\epsilon = \mathrm{diag}((-1)^{n-n_r}I_1, I_{2n_r}), \det(\epsilon) = (-1)^{n-n_r}$. Let $C_r' = \epsilon C_r$. We see that

$$\begin{aligned} V_r &\cong \Big\{(A_r^1, B_r^1, \dots, A_r^\ell, B_r^\ell, C_r') \in SO(2n_r+1)^{2\ell} \times O(2n_r+1) \mid \\ &\qquad \prod_{i=1}^{\ell}[A_r^i, B_r^i] = (C_r')^2, \det(C_r') = (-1)^{n-n_r}\Big\} \\ &\cong V^{\ell,1}_{O(2n_r+1),(-1)^{n-n_r}} \end{aligned}$$

where $V^{\ell,1}_{O(n),\pm1}$ is the twisted representation variety defined in (4.11) of Section 4.7. $V^{\ell,1}_{O(n),\pm1}$ is nonempty if $\ell \geq 2$. We have shown that $V^{\ell,1}_{O(n),\pm1}$ is disconnected with two components $V^{\ell,1,+1}_{O(n),\pm1}$ and $V^{\ell,1,-1}_{O(n),\pm1}$ if $\ell \geq 2$ and $n > 2$ (Proposition 4.14).

We have

$$V^{\ell,1}_{\mathrm{YM}}(SO(2n+1))_\mu = \prod_{j=1}^{r} V_j.$$

We define a $U(n_j)$-action on $V_j = \tilde{V}^{\ell,1}_{n_j,-k_j}$ by (4.9) of Section 4.6; when $k_r = 0$, we define an $SO(2n_r+1)$-action on $V_r = V^{\ell,1}_{O(2n_r+1),(-1)^{n-n_r}}$ by (4.13) of Section 4.7

Then we have a homeomorphism

$$V^{\ell,1}_{\mathrm{YM}}(SO(2n+1))_\mu/SO(2n+1)_{X_\mu} \cong \begin{cases} \prod_{j=1}^{r}(V_j/U(n_j)), & k_r>0,\\ \prod_{j=1}^{r-1}(V_j/U(n_j))\times V_r/SO(2n_r+1), & k_r=0,\end{cases}$$

and a homotopy equivalence

$$V^{\ell,1}_{\mathrm{YM}}(SO(2n+1))_\mu{}^{hSO(2n+1)_{X_\mu}} \sim \begin{cases} \prod_{j=1}^{r} V_j{}^{hU(n_j)}, & k_r>0,\\ \prod_{j=1}^{r-1} V_j{}^{hU(n_j)}\times V_r{}^{hSO(2n_r+1)}. & k_r=0,\end{cases}$$

$i=2$. Let $\epsilon_j=\mathrm{diag}(H_{n_j})$. Define

$$\begin{aligned} V_j &= \Big\{(A^1_j,B^1_j,\ldots,A^\ell_j,B^\ell_j,D_j,C_j)\in\Phi(U(n_j))^{2\ell+2}\mid \\ &\qquad \prod_{i=1}^{\ell}[A^i_j,B^i_j]=T_{n_j,k_j}\epsilon_jC_jD_j(\epsilon_jC_j)^{-1}D_j\Big\}\\ &=\Big\{(A^1_j,B^1_j,\ldots,A^\ell_j,B^\ell_j,D_j,C_j)\in\Phi(U(n_j))^{2\ell+2}\mid \\ &\qquad \prod_{i=1}^{\ell}[A^i_j,B^i_j]=T_{n_j,k_j}\epsilon_jC_j\epsilon_j^{-1}\epsilon_jD_j\epsilon^{-1}\epsilon_jC_j^{-1}\epsilon_j^{-1}D_j\Big\}\\ &\overset{\Phi}{\cong}\Big\{(A^1_j,B^1_j,\ldots,A^\ell_j,B^\ell_j,D_j,C_j)\in U(n_j)^{2\ell+2}\mid \\ &\qquad \prod_{i=1}^{\ell}[A^i_j,B^i_j]=e^{2\pi\sqrt{-1}k_j/n_j}\bar{C}_j\bar{D}_j\bar{C}_j^{-1}D_j\Big\}\cong\tilde{V}^{\ell,2}_{n_j,-k_j}\end{aligned}\tag{5.9}$$

where $\tilde{V}^{\ell,2}_{n_j,-k_j}$ is the twisted representation variety defined in (4.8) of Section 4.6. $\tilde{V}^{\ell,2}_{n_j,-k_j}$ is nonempty if $\ell\geq 1$. We have shown that $\tilde{V}^{\ell,2}_{n_j,-k_j}$ is connected if $\ell\geq 4$ (Proposition 4.13).

When $k_r>0$, define V_r by (5.9). When $k_r=0$, define

$$\begin{aligned} V_r=&\Big\{(A^1_r,B^1_r,\ldots,A^\ell_r,B^\ell_r,D_r,C_r)\in SO(2n_r+1)^{2\ell+2}\mid\\ &\prod_{i=1}^{\ell}[A^i_r,B^i_r]=\epsilon C_rD_r(\epsilon C_r)^{-1}D_r\Big\},\end{aligned}\tag{5.10}$$

where $\epsilon=\mathrm{diag}((-1)^{n-n_r}I_1,I_{2n_r}),\det(\epsilon)=(-1)^{n-n_r}$. Let $C'_r=\epsilon C_r$. We see that

$$\begin{aligned} V_r &\cong \Big\{(A^1_r,B^1_r,\ldots,A^\ell_r,B^\ell_r,D_r,C'_r)\in SO(2n_r+1)^{2\ell+1}\times O(2n_r+1)\mid\\ &\qquad\prod_{i=1}^{\ell}[A^i_r,B^i_r]=C'_rD_rC'^{-1}_rD_r,\det(C'_r)=(-1)^{n-n_r}\Big\}\\ &\cong V^{\ell,2}_{O(2n_r+1),(-1)^{n-n_r}}\end{aligned}$$

where $V^{\ell,2}_{O(n),\pm1}$ is the twisted representation variety defined in (4.12) of Section 4.7. $V^{\ell,2}_{O(n),\pm1}$ is nonempty if $\ell\geq 4$. We have shown that $V^{\ell,2}_{O(n),\pm1}$ is disconnected with two components $V^{\ell,2,+1}_{O(n),\pm1}$ and $V^{\ell,2,-1}_{O(n),\pm1}$ if $\ell\geq 4$ and $n>2$ (Proposition 4.14).

We have

$$V^{\ell,2}_{\mathrm{YM}}(SO(2n+1))_\mu = \prod_{j=1}^{r} V_j.$$

We define a $U(n_j)$-action on $V_j = \tilde{V}^{\ell,2}_{n_j,-k_j}$ by (4.10) of Section 4.6; when $k_r = 0$, we define an $SO(2n_r+1)$-action on $V_r = V^{\ell,2}_{O(2n_r+1),(-1)^{n-n_r}}$ by (4.14) of Section 4.7. Then we have a homeomorphism

$$V^{\ell,2}_{\mathrm{YM}}(SO(2n+1))_\mu/SO(2n+1)_{X_\mu} \cong \begin{cases} \prod_{j=1}^{r}(V_j/U(n_j)), & k_r > 0, \\ \prod_{j=1}^{r-1}(V_j/U(n_j)) \times V_r/SO(2n_r+1), & k_r = 0, \end{cases}$$

and a homotopy equivalence

$$V^{\ell,2}_{\mathrm{YM}}(SO(2n+1))_\mu{}^{hSO(2n+1)_{X_\mu}} \sim \begin{cases} \prod_{j=1}^{r} V_j{}^{hU(n_j)}, & k_r > 0, \\ \prod_{j=1}^{r-1} V_j{}^{hU(n_j)} \times V_r{}^{hSO(2n_r+1)}, & k_r = 0, \end{cases}$$

We have seen that for $i = 1, 2$, $V^{\ell,i}_{\mathrm{YM}}(SO(2n+1))_\mu$ is connected when $k_r > 0$. In this case, to determine the topological type of the underlying $SO(2n+1)$-bundle P over Σ^ℓ_i, we can just look at a special point in $V^{\ell,i}_{\mathrm{YM}}(SO(2n+1))_\mu$ where c', d are the identity element I_{2n+1}. Then

$$\prod_{i=1}^{\ell}[a_i, b_i] = \exp(X_\mu/2),$$

so $a_1, b_1, \ldots, a_\ell, b_\ell$ can be viewed as the holonomies of a Yang-Mills connection on an $SO(2n+1)$-bundle $Q_0 \to \Sigma^\ell_0$. Also, $c = \epsilon = \mathrm{diag}(H_n, (-1)^n I_1)$ can be viewed as the holonomy of a flat connection on an $SO(2n+1)$-bundle Q_1 over $\Sigma^0_1 = \mathbb{RP}^2$, and $c = \epsilon$, $d = I_{2n+1}$ can be viewed as the holonomies of a flat connection on an $SO(2n+1)$-bundle Q_2 over Σ^0_2 (a Klein bottle). Let Σ' be obtained by gluing Σ^ℓ_0 and Σ^0_i at a point, and let $P' \to \Sigma'$ be the (topological) principal $SO(2n+1)$-bundle over Σ' such that $P'|_{\Sigma^\ell_0} = Q_0$ and $P'|_{\Sigma^0_i} = Q_i$. Then $P = p^*P'$ where $p : \Sigma^\ell_i \to \Sigma' = \Sigma^\ell_0 \cup \Sigma^0_i$ is the collapsing map. Then $w_2(P') = (w_2(Q_0), w_2(Q_i))$ under the isomorphism

$$H^2(\Sigma'; \mathbb{Z}/2\mathbb{Z}) \cong H^2(\Sigma^\ell_0; \mathbb{Z}/2\mathbb{Z}) \oplus H^2(\Sigma^0_i; \mathbb{Z}/2\mathbb{Z}),$$

and $w_2(P) = p^*w_2(P') = w_2(Q_0) + w_2(Q_i)$, if we identify $H^2(\Sigma^\ell_i; \mathbb{Z}/2\mathbb{Z})$, $H^2(\Sigma^\ell_0; \mathbb{Z}/2\mathbb{Z})$, and $H^2(\Sigma^0_i; \mathbb{Z}/2\mathbb{Z})$ with $\mathbb{Z}/2\mathbb{Z}$. So it remains to compute $w_2(Q_0)$, $w_2(Q_1)$, and $w_2(Q_2)$. We have $Q_0 \cong P^{n,-(k_1+\cdots+k_r)} \times_{U(n)} SO(2n+1)$, so $w_2(Q_0) = k_1 + \cdots + k_r$ (mod 2). To compute $w_2(Q_1)$ and $w_2(Q_2)$, we lift $c = \epsilon$ to $\tilde{c} \in Spin(2n+1)$ and lift $d = I_{2n+1}$ to $\tilde{d} \in Spin(2n+1)$. Since $2\pi_1(SO(2n+1))$ is the trivial group, we may choose any lifting for c and d. We choose $\tilde{d} = 1 \in Spin(2n+1)$ and

$$\tilde{c} = \begin{cases} e_2 e_4 \cdots e_{2n}, & n \text{ even}, \\ e_2 e_4 \cdots e_{2n} e_{2n+1}, & n \text{ odd}. \end{cases}$$

Then $\tilde{c}^2 = (-1)^{n(n+1)/2}$ and $\tilde{c}\tilde{d}\tilde{c}^{-1}\tilde{d} = 1$. We conclude that

$$w_2(Q_1) = \frac{n(n+1)}{2} \pmod 2, \quad w_2(Q_2) = 0 \pmod 2,$$

so

$$w_2(P) = k_1 + \cdots + k_r + i\frac{n(n+1)}{2} \pmod 2.$$

When $k_r = 0$, we have seen that $V^{\ell,i}_{\mathrm{YM}}(SO(2n+1))_\mu$ is disconnected with two components. To determine the corresponding underlying topological types, we consider two special cases.

Case 1. We consider special points

$$(a_1, b_1, \ldots, a_\ell, b_\ell, c) \in V^{\ell,1}_{\mathrm{YM}}(SO(2n+1))_\mu,\ (a_1, b_1, \ldots, a_\ell, b_\ell, d, c) \in V^{\ell,2}_{\mathrm{YM}}(SO(2n+1))_\mu,$$

where

$$\begin{aligned} a_i &= \mathrm{diag}(A^i_1, \ldots, A^i_{r-1}, I_{2n_r+1}), \quad b_i = \mathrm{diag}(B^i_1, \ldots, B^i_{r-1}, I_{2n_r+1}), \\ c &= \epsilon_\mu = \mathrm{diag}(H_{n-n_r}, (-1)^{n-n_r} I_1, I_{2n_r}), \quad d = I_{2n+1}. \end{aligned}$$

Let $\epsilon_1 = \mathrm{diag}((-1)^{n-n_r} I_1, I_{2n_r})$. Then

$$\begin{aligned} &(A^i_j, B^i_j, \ldots, A^i_j, B^i_j) \in X^{\ell,0}_{\mathrm{YM}}(U(n_j))_{-\frac{k_j}{n_j},\ldots,-\frac{k_j}{n_j}}, \quad j = 1, \ldots, r-1, \\ &(I_{2n_r+1}, \ldots, I_{2n_r+1}, \epsilon_1) \in V^{\ell,1,(-1)^{n-n_r}}_{O(2n_r+1),(-1)^{n-n_r}}, \\ &(I_{2n_r+1}, \ldots, I_{2n_r+1}, I_{2n_r+1}, \epsilon_1) \in V^{\ell,2,1}_{O(2n_r+1),(-1)^{n-n_r}}. \end{aligned}$$

We have $P = P_1 \times P_2$, where P_1 is an $SO(2(n-n_r)+1)$-bundle, and P_2 is an $SO(2n_r)$-bundle with trivial holonomies I_{2n_r}. We have

$$w_2(P) = w_2(P_1) = k_1 + \cdots + k_{r-1} + i\frac{(n-n_r)(n-n_r+1)}{2}.$$

Case 2. We consider special points

$$(a_1, b_1, \ldots, a_\ell, b_\ell, c) \in V^{\ell,1}_{\mathrm{YM}}(SO(2n+1))_\mu,\ (a_1, b_1, \ldots, a_\ell, b_\ell, d, c) \in V^{\ell,2}_{\mathrm{YM}}(SO(2n+1))_\mu,$$

where

$$\begin{aligned} a_i &= \mathrm{diag}(A^i_1, \ldots, A^i_{r-1}, I_{2n_r+1}), \quad b_i = \mathrm{diag}(B^i_1, \ldots, B^i_{r-1}, I_{2n_r+1}), \\ c &= \mathrm{diag}(H_{n-n_r}, (-1)^{(n-n_r)} I_1, -I_2, I_{2n_r-2}), \quad d = \mathrm{diag}(I_{2(n-n_r)+1}, -I_2, I_{2n_r-2}). \end{aligned}$$

Let $\epsilon_1 = \mathrm{diag}((-1)^{n-n_r} I_1, -I_2, I_{2n_r-2})$, $\epsilon_2 = \mathrm{diag}(I_1, -I_2, I_{2n_r-2})$. Then

$$\begin{aligned} &(A^i_j, B^i_j, \ldots, A^i_j, B^i_j) \in X^{\ell,0}_{\mathrm{YM}}(U(n_j))_{-\frac{k_j}{n_j},\ldots,-\frac{k_j}{n_j}}, \quad j = 1, \ldots, r-1, \\ &(I_{2n_r+1}, \ldots, I_{2n_r+1}, \epsilon_1) \in V^{\ell,1,-(-1)^{n-n_r}}_{O(2n_r+1),(-1)^{n-n_r}}, \\ &(I_{2n_r+1}, \ldots, I_{2n_r+1}, \epsilon_2, \epsilon_1) \in V^{\ell,2,-1}_{O(2n_r+1),(-1)^{n-n_r}}. \end{aligned}$$

We have $P = P_1 \times P_2$, where P_1 is an $SO(2(n-n_r)+1)$-bundle, and P_2 is an $SO(2n_r)$-bundle with holonomies $a_i = b_i = I_{2n_r}$, $c = d = \epsilon = \mathrm{diag}(-I_2, I_{2n_r-2})$. Similarly, we can choose the lifting of d and c as $\tilde{d} = \tilde{c} = e_1 e_2$. Then $\tilde{c}^2 = \tilde{c}\tilde{d}\tilde{c}^{-1}\tilde{d} = -1$. We have

$$w_2(P_1) = k_1 + \cdots + k_{r-1} + i\frac{(n-n_r)(n-n_r+1)}{2} \pmod 2, \quad w_2(P_2) = 1 \pmod 2,$$

so

$$w_2(P) = w_2(P_1) + w_2(P_2) = k_1 + \cdots + k_{r-1} + i\frac{(n-n_r)(n-n_r+1)}{2} + 1 \pmod 2.$$

To summarize, when $k_r = 0$ we have

$$V^{\ell,i}_{\mathrm{YM}}(SO(2n+1))^{\pm}_{\mu} = \prod_{j=1}^{r-1} V_j \times V^{\ell,i,\pm(-1)^{k_1+\cdots+k_{r-1}+i\frac{(n-n_r)(n-n_r-1)}{2}}}_{O(2n_r+1),(-1)^{n-n_r}},$$

where $V^{\ell,i}_{\mathrm{YM}}(SO(2n+1))^{\pm}_{\mu}$ is the ϵ_μ-reduced version of $X^{\ell,i}_{\mathrm{YM}}(SO(2n+1))^{\pm 1}_{\mu}$.

To simplify the notation, we write

$$\mu = (\mu_1, \ldots, \mu_n) = \Big(\underbrace{\frac{2k_1}{n_1}, \ldots, \frac{2k_1}{n_1}}_{n_1}, \ldots, \underbrace{\frac{2k_r}{n_r}, \ldots, \frac{2k_r}{n_r}}_{n_r}\Big)$$

instead of

$$\sqrt{-1}\mathrm{diag}\Big(\frac{2k_1}{n_1}J_{n_1}, \ldots, \frac{2k_r}{n_r}J_{n_r}, 0I_1\Big).$$

Let

$$\begin{aligned}
\hat{I}_{SO(2n+1)} &= \Big\{\mu = \Big(\underbrace{\frac{2k_1}{n_1}, \ldots, \frac{2k_1}{n_1}}_{n_1}, \ldots, \underbrace{\frac{2k_r}{n_r}, \ldots, \frac{2k_r}{n_r}}_{n_r}\Big) \mid n_j \in \mathbb{Z}_{>0}, \\
&\qquad n_1 + \cdots + n_r = n,\ k_j \in \mathbb{Z},\ \frac{k_1}{n_1} > \cdots > \frac{k_r}{n_r} \geq 0\Big\}, \\
\hat{I}^{\pm 1}_{SO(2n+1)} &= \{\mu \in \hat{I}_{SO(2n+1)} \mid \mu_n > 0, (-1)^{k_1+\cdots+k_r+\frac{in(n+1)}{2}} = \pm 1\}, \\
\hat{I}^{0}_{SO(2n+1)} &= \{\mu \in \hat{I}_{SO(2n+1)} \mid \mu_n = 0\}.
\end{aligned}$$

For $i = 1, 2$, define twisted moduli spaces

$$\tilde{\mathcal{M}}^{\ell,i}_{n,k} = \tilde{V}^{\ell,i}_{n,k}/U(n), \quad \mathcal{M}^{\ell,i,\pm 1}_{O(n),\pm 1} = V^{\ell,i,\pm 1}_{O(n),\pm 1}/SO(n).$$

PROPOSITION 5.9. *Suppose that $\ell \geq 2i$, where $i = 1, 2$. Let*

$$\mu = \Big(\underbrace{\frac{2k_1}{n_1}, \ldots, \frac{2k_1}{n_1}}_{n_1}, \ldots, \underbrace{\frac{2k_r}{n_r}, \ldots, \frac{2k_r}{n_r}}_{n_r}\Big) \in \hat{I}_{SO(2n+1)}. \tag{5.11}$$

(i) *If $\mu \in \hat{I}^{\pm 1}_{SO(2n+1)}$, then $X^{\ell,i}_{\mathrm{YM}}(SO(2n+1))_{\mu} = X^{\ell,i}_{\mathrm{YM}}(SO(2n+1))^{\pm 1}_{\mu}$ is nonempty and connected (coming from either the trivial bundle or the nontrivial bundle). We have a homeomorphism*

$$X^{\ell,i}_{\mathrm{YM}}(SO(2n+1))_{\mu}/SO(2n+1) \cong \prod_{j=1}^{r} \tilde{\mathcal{M}}^{\ell,i}_{n_j,-k_j}$$

and a homotopy equivalence

$$X^{\ell,i}_{\mathrm{YM}}(SO(2n+1))_{\mu}^{hSO(2n+1)} \sim \prod_{j=1}^{r} (\tilde{V}^{\ell,i}_{n_j,-k_j})^{hU(n_j)}.$$

(ii) *If $\mu \in \hat{I}^{0}_{SO(2n+1)}$, then $X^{\ell,i}_{\mathrm{YM}}(SO(2n+1))_{\mu}$ has two connected components (coming from both bundles)*

$$X^{\ell,i}_{\mathrm{YM}}(SO(2n+1))^{+1}_{\mu} \quad \text{and} \quad X^{\ell,i}_{\mathrm{YM}}(SO(2n+1))^{-1}_{\mu}.$$

We have homeomorphisms

$$X^{\ell,i}_{\mathrm{YM}}(SO(2n+1))^{\pm 1}_{\mu}/SO(2n+1)$$
$$\cong \prod_{j=1}^{r-1}\tilde{\mathcal{M}}^{\ell,i}_{n_j,-k_j}\times \mathcal{M}^{\ell,i,\pm(-1)^{k_1+\cdots+k_{r-1}+i\frac{(n-n_r)(n-n_r-1)}{2}}}_{O(2n_r+1),(-1)^{n-n_r}}$$

and homotopy equivalences

$$\left(X^{\ell,i}_{\mathrm{YM}}(SO(2n+1))^{\pm 1}_{\mu}\right)^{hSO(2n+1)}$$
$$\sim \prod_{j=1}^{r-1}(\tilde{V}^{\ell,i}_{n_j,-k_j})^{hU(n_j)}\times\left(V^{\ell,i,\pm(-1)^{k_1+\cdots+k_{r-1}+i\frac{(n-n_r)(n-n_r-1)}{2}}}_{O(2n_r+1),(-1)^{n-n_r}}\right)^{hSO(2n_r+1)}.$$

PROPOSITION 5.10. *Suppose that $\ell\geq 2i$, where $i=0,1$. The connected components of $X^{\ell,i}_{\mathrm{YM}}(SO(2n+1))^{\pm1}$ are*

$$\{X^{\ell,i}_{\mathrm{YM}}(SO(2n+1))_{\mu}\mid \mu\in\hat{I}^{\pm1}_{SO(2n+1)}\}\cup\{X^{\ell,i}_{\mathrm{YM}}(SO(2n+1))^{\pm1}_{\mu}\mid\mu\in\hat{I}^{0}_{SO(2n+1)}\}.$$

Notice that, the set $\{\mu=\sqrt{-1}\mathrm{diag}(\mu_1 J,\ldots,\mu_n J,0I_1)\mid(\mu_1,\ldots,\mu_n)\in\hat{I}_{SO(2n+1)}\}$ is a *proper* subset of $\{\mu\in(\Xi^I_+)^\tau|I\subseteq\Delta,\tau(I)=I\}$ as mentioned in Section 4.5.

The following is an immediate consequence of Proposition 5.9.

THEOREM 5.11. *Suppose that $\ell\geq 2i$, where $i=1,2$, and let μ be as in (5.11).*

(i) *If $\mu\in\hat{I}^{\pm1}_{SO(2n+1)}$, then*

$$P^{SO(2n+1)}_t\left(X^{\ell,0}_{\mathrm{YM}}(SO(2n+1))_{\mu}\right)=\prod_{j=1}^{r}P^{U(n_j)}_t(\tilde{V}^{\ell,i}_{n_j,-k_j}).$$

(ii) *If $\mu\in\hat{I}^{0}_{SO(2n+1)}$, then*

$$P^{SO(2n+1)}_t\left(X^{\ell,i}_{\mathrm{YM}}(SO(2n+1))^{\pm1}_{\mu}\right)$$
$$=\prod_{j=1}^{r-1}P^{U(n_j)}_t(\tilde{V}^{\ell,i}_{n_j,-k_j})\cdot P^{SO(2n_r+1)}_t\left(V^{\ell,i,\pm(-1)^{k_1+\cdots+k_{r-1}+i\frac{(n-n_r)(n-n_r-1)}{2}}}_{O(2n_r+1),(-1)^{n-n_r}}\right).$$

CHAPTER 6

Yang-Mills $SO(2n)$-Connections

The maximal torus of $SO(2n)$ consists of block diagonal matrices of the form

$$\mathrm{diag}(A_1, \ldots, A_n)$$

where $A_1, \ldots, A_n \in SO(2)$. The Lie algebra of the maximal torus consists of matrices of the form

$$2\pi \mathrm{diag}(t_1 J, \ldots, t_n J)$$

where

$$J = \begin{pmatrix} 0 & -1 \\ 1 & 0 \end{pmatrix}.$$

The fundamental Weyl chamber is

$$\overline{C}_0 = \{\sqrt{-1}\mathrm{diag}(t_1 J, \ldots, t_n J) \mid t_1 \geq t_2 \geq \cdots \geq | t_n | \geq 0\}.$$

As in Chapter 5, in this chapter we continue to assume

$$n_1, \ldots, n_r \in \mathbb{Z}_{>0}, \quad n_1 + \cdots + n_r = n.$$

6.1. $SO(2n)$-connections on orientable surfaces

There are four cases.

Case 1. $t_{n-1} > |t_n|$, $n_r = 1$.

$$\mu = \sqrt{-1}\mathrm{diag}(\lambda_1 J_{n_1}, \ldots, \lambda_{r-1} J_{n_{r-1}}, \lambda_r J),$$

where $\lambda_1 > \cdots > \lambda_{r-1} > |\lambda_r| \geq 0$. Thus

$$SO(2n)_{X_\mu} \cong \Phi(U(n_1)) \times \cdots \times \Phi(U(n_{r-1})) \times \Phi(U(n_r)).$$

Suppose that $(a_1, b_1, \ldots, a_\ell, b_\ell, X_\mu) \in X^{\ell,0}_{\mathrm{YM}}(SO(2n))$. Then

$$\exp(X_\mu) = \prod_{i=1}^{\ell} [a_i, b_i]$$

where $a_1, b_1, \ldots, a_\ell, b_\ell \in SO(2n)_{X_\mu}$. Then we have

$$\exp(X_\mu) \in (SO(2n)_{X_\mu})_{ss} = \Phi(SU(n_1)) \times \cdots \times \Phi(SU(n_{r-1})) \times \{I_2\}.$$

Thus

$$\begin{aligned} X_\mu &= 2\pi \mathrm{diag}\Big(\frac{k_1}{n_1} J_{n_1}, \ldots, \frac{k_{r-1}}{n_{r-1}} J_{n_{r-1}}, k_r J\Big), \\ \mu &= \sqrt{-1}\mathrm{diag}\Big(\frac{k_1}{n_1} J_{n_1}, \ldots, \frac{k_{r-1}}{n_{r-1}} J_{n_{r-1}}, k_r J\Big), \end{aligned}$$

where

$$k_1, \ldots, k_r \in \mathbb{Z}, \quad \frac{k_1}{n_1} > \cdots > \frac{k_{r-1}}{n_{r-1}} > |k_r| \geq 0.$$

Recall that for each μ, the representation variety is

$$V^{\ell,0}_{\mathrm{YM}}(SO(2n))_\mu = \{(a_1, b_1, \dots, a_\ell, b_\ell) \in (SO(2n)_{X_\mu})^{2\ell} \mid \prod_{i=1}^{\ell}[a_i, b_i] = \exp(X_\mu)\}.$$

For $i = 1, \dots, \ell$, write

$$a_i = \mathrm{diag}(A^i_1, \dots, A^i_r), \quad b_i = \mathrm{diag}(B^i_1, \dots, B^i_r),$$

where $A^i_j,\ B^i_j \in \Phi(U(n_j))$. Define V_j as in (5.4). Then

$$V^{\ell,0}_{\mathrm{YM}}(SO(2n))_\mu = \prod_{j=1}^{r} V_j. \tag{6.1}$$

We have a homeomorphism

$$V^{\ell,0}_{\mathrm{YM}}(SO(2n))_\mu / SO(2n)_{X_\mu} \cong \prod_{j=1}^{r}(V_j/U(n_j)) \tag{6.2}$$

and a homotopy equivalence

$$V^{\ell,0}_{\mathrm{YM}}(SO(2n))_\mu{}^{hSO(2n)_{X_\mu}} \sim \prod_{j=1}^{r} V_j{}^{hU(n_j)}. \tag{6.3}$$

Case 2. $t_{n-1} = -t_n > 0$, $n_r > 1$.

$$\mu = \sqrt{-1}\mathrm{diag}(\lambda_1 J_{n_1}, \dots, \lambda_{r-1} J_{n_{r-1}}, \lambda_r J_{n_r - 1}, -\lambda_r J),$$

where $\lambda_1 > \cdots > \lambda_r > 0$. Thus

$$SO(2n)_{X_\mu} \cong \Phi(U(n_1)) \times \cdots \times \Phi(U(n_{r-1})) \times \Phi'(U(n_r)),$$

where $\Phi' : U(m) \hookrightarrow SO(2m)$ is the embedding defined as follows. Consider the $\mathbb{R}$-linear map $L' : \mathbb{R}^{2m} \to \mathbb{C}^m$ given by

$$\begin{pmatrix} x_1 \\ y_1 \\ \vdots \\ x_{m-1} \\ y_{m-1} \\ x_m \\ y_m \end{pmatrix} \mapsto \begin{pmatrix} x_1 + \sqrt{-1}y_1 \\ \vdots \\ x_{m-1} + \sqrt{-1}y_{m-1} \\ x_m - \sqrt{-1}y_m \end{pmatrix}.$$

We have $(L')^{-1} \circ \sqrt{-1}I_m \circ L'(v) = \mathrm{diag}(J_{m-1}, -J)(v)$ for $v \in \mathbb{R}^{2m}$. If A is a $m \times m$ matrix, and let $\Phi'(A)$ be the $(2m) \times (2m)$ matrix given by

$$(L')^{-1} \circ A \circ L'(v) = \Phi'(A)(v).$$

Note that $A(\sqrt{-1}I_m) = (\sqrt{-1}I_m)A \Rightarrow \Phi'(A)\mathrm{diag}(J_{m-1}, -J) = \mathrm{diag}(J_{m-1}, -J)\Phi'(A)$.

Suppose that $(a_1, b_1, \dots, a_\ell, b_\ell, X_\mu) \in X^{\ell,0}_{\mathrm{YM}}(SO(2n))$. Then

$$\exp(X_\mu) = \prod_{i=1}^{\ell}[a_i, b_i]$$

where $a_1, b_1, \dots, a_\ell, b_\ell \in SO(2n)_{X_\mu}$. Then we have

$$\exp(X_\mu) \in (SO(2n)_{X_\mu})_{ss} = \Phi(SU(n_1)) \times \cdots \times \Phi(SU(n_{r-1})) \times \Phi'(SU(n_r)).$$

Thus

$$\begin{aligned} X_\mu &= 2\pi\mathrm{diag}\Big(\frac{k_1}{n_1}J_{n_1},\ldots,\frac{k_{r-1}}{n_{r-1}}J_{n_{r-1}},\frac{k_r}{n_r}J_{n_r-1},\frac{-k_r}{n_r}J\Big),\\ \mu &= \sqrt{-1}\mathrm{diag}\Big(\frac{k_1}{n_1}J_{n_1},\ldots,\frac{k_{r-1}}{n_{r-1}}J_{n_{r-1}},\frac{k_r}{n_r}J_{n_r-1},\frac{-k_r}{n_r}J\Big), \end{aligned}$$

where

$$k_1,\ldots,k_r\in\mathbb{Z},\qquad \frac{k_1}{n_1}>\cdots>\frac{k_r}{n_r}>0.$$

Recall that for each μ, the representation variety is

$$V^{\ell,0}_{\mathrm{YM}}(SO(2n))_\mu=\{(a_1,b_1,\ldots,a_\ell,b_\ell)\in(SO(2n)_{X_\mu})^{2\ell}\mid \prod_{i=1}^{\ell}[a_i,b_i]=\exp(X_\mu)\}.$$

For $i=1,\ldots,\ell$, write

$$a_i=\mathrm{diag}(A^i_1,\ldots,A^i_r),\quad b_i=\mathrm{diag}(B^i_1,\ldots,B^i_r),$$

where $A^i_j,\ B^i_j\in\Phi(U(n_j))$ for $j=1,\ldots,r-1$, and $A^i_r,\ B^i_r\in\Phi'(U(n_r))$.
For $j=1,\ldots,r-1$, define V_j as in (5.4). Recall that

$$\hat{J}_t=\begin{pmatrix}\cos(2\pi t) & -\sin(2\pi t)\\ \sin(2\pi t) & \cos(2\pi t)\end{pmatrix}.$$

Define

$$\begin{aligned} V_r &= \Big\{(A^1_r,B^1_r,\ldots,A^\ell_r,B^\ell_r)\in\Phi'(U(n_r))^{2\ell}\mid\\ &\qquad \prod_{i=1}^{\ell}[A^i_r,B^i_r]=\mathrm{diag}(\hat{J}_{k_r/n_r},\ldots,\hat{J}_{k_r/n_r},\hat{J}_{-k_r/n_r})\Big\}\\ &\overset{\Phi'}{\cong} \Big\{(A^1_r,B^1_r,\ldots,A^\ell_r,B^\ell_r)\in U(n_r)^{2\ell}\mid \prod_{i=1}^{\ell}[A^i_r,B^i_r]=e^{2\pi\sqrt{-1}k_r/n_r}I_{n_r}\Big\}\\ &\cong X^{\ell,0}_{\mathrm{YM}}(U(n_r))_{-\frac{k_r}{n_r},\ldots,-\frac{k_r}{n_r}}. \end{aligned}$$

Then we have (6.1), (6.2), and (6.3).

Case 3. $t_{n-1}=t_n>0$, $n_r>1$.

$$\mu=\sqrt{-1}\mathrm{diag}(\lambda_1J_{n_1},\ldots,\lambda_rJ_{n_r}),$$

where $\lambda_1>\cdots>\lambda_r>0$. Thus

$$SO(2n)_{X_\mu}\cong\Phi(U(n_1))\times\cdots\times\Phi(U(n_r)).$$

Let $X_\mu=-2\pi\sqrt{-1}\mu$. Suppose that $(a_1,b_1,\ldots,a_\ell,b_\ell,X_\mu)\in X^{\ell,0}_{\mathrm{YM}}(SO(2n))$. Then

$$\exp(X_\mu)=\prod_{i=1}^{\ell}[a_i,b_i]$$

where $a_1,b_1,\ldots,a_\ell,b_\ell\in SO(2n)_{X_\mu}$. Then we have

$$\exp(X_\mu)\in(SO(2n)_{X_\mu})_{ss}=\Phi(SU(n_1))\times\cdots\times\Phi(SU(n_r)).$$

Thus

$$\begin{aligned} X_\mu &= 2\pi \mathrm{diag}\Big(\frac{k_1}{n_1}J_{n_1},\ldots,\frac{k_r}{n_r}J_{n_r}\Big),\\ \mu &= \sqrt{-1}\mathrm{diag}\Big(\frac{k_1}{n_1}J_{n_1},\ldots,\frac{k_r}{n_r}J_{n_r}\Big), \end{aligned}$$

where

$$k_1,\ldots,k_r\in\mathbb{Z},\quad \frac{k_1}{n_1}>\cdots>\frac{k_r}{n_r}>0.$$

Recall that for each μ, the representation variety is

$$V^{\ell,0}_{\mathrm{YM}}(SO(2n))_\mu=\{(a_1,b_1,\ldots,a_\ell,b_\ell)\in(SO(2n)_{X_\mu})^{2\ell}\mid\prod_{i=1}^{\ell}[a_i,b_i]=\exp(X_\mu)\}.$$

For $i=1,\ldots,\ell$, write

$$a_i=\mathrm{diag}(A^i_1,\ldots,A^i_r),\quad b_i=\mathrm{diag}(B^i_1,\ldots,B^i_r),$$

where $A^i_j,\ B^i_j\in\Phi(U(n_j))$.

Define V_j as in (5.4). Then we have (6.1), (6.2), and (6.3).

Case 4. $t_{n-1}=t_n=0$**,** $n_r>1$**.**

$$\mu=\sqrt{-1}\mathrm{diag}(\lambda_1 J_{n_1},\ldots,\lambda_{r-1}J_{n_{r-1}},0J_{n_r}),$$

where $\lambda_1>\cdots>\lambda_{r-1}>0$. Thus

$$SO(2n)_{X_\mu}\cong\Phi(U(n_1))\times\cdots\times\Phi(U(n_{r-1}))\times SO(2n_r).$$

Let $X_\mu=-2\pi\sqrt{-1}\mu$. Suppose that $(a_1,b_1,\ldots,a_\ell,b_\ell,X_\mu)\in X^{\ell,0}_{\mathrm{YM}}(SO(2n))$. Then

$$\exp(X_\mu)=\prod_{i=1}^{\ell}[a_i,b_i]$$

where $a_1,b_1,\ldots,a_\ell,b_\ell\in SO(2n)_{X_\mu}$. Then we have

$$\exp(X_\mu)\in(SO(2n)_{X_\mu})_{ss}=\Phi(SU(n_1))\times\cdots\times\Phi(SU(n_{r-1}))\times SO(2n_r).$$

Thus

$$\begin{aligned} X_\mu &= 2\pi \mathrm{diag}\Big(\frac{k_1}{n_1}J_{n_1},\ldots,\frac{k_{r-1}}{n_{r-1}}J_{n_{r-1}},0J_{n_r}\Big),\\ \mu &= \sqrt{-1}\mathrm{diag}\Big(\frac{k_1}{n_1}J_{n_1},\ldots,\frac{k_{r-1}}{n_{r-1}}J_{n_{r-1}},0J_{n_r}\Big), \end{aligned}$$

where

$$k_1,\ldots,k_{r-1}\in\mathbb{Z},\quad \frac{k_1}{n_1}>\cdots>\frac{k_{r-1}}{n_{r-1}}>0.$$

Recall that for each μ, the representation variety is

$$V^{\ell,0}_{\mathrm{YM}}(SO(2n))_\mu=\{(a_1,b_1,\ldots,a_\ell,b_\ell)\in(SO(2n)_{X_\mu})^{2\ell}\mid\prod_{i=1}^{\ell}[a_i,b_i]=\exp(X_\mu)\}.$$

For $i=1,\ldots,\ell$, write

$$a_i=\mathrm{diag}(A^i_1,\ldots,A^i_r),\quad b_i=\mathrm{diag}(B^i_1,\ldots,B^i_r),$$

where $A^i_j,\ B^i_j\in\Phi(U(n_j))$ for $j=1,\ldots,r-1$, and $A^i_r,\ B^i_r\in SO(2n_r)$.

For $j = 1, \ldots, r-1$, define V_j as in (5.4). Define

$$V_r = \Big\{(A_r^1, B_r^1, \ldots, A_r^\ell, B_r^\ell) \in SO(2n_r)^{2\ell} \mid \prod_{i=1}^{\ell}[A_r^i, B_r^i] = I_{2n_r}\Big\} \cong X_{\text{flat}}^{\ell,0}(SO(2n_r)).$$

Then $V_{\text{YM}}^{\ell,0}(SO(2n))_\mu = \prod_{j=1}^r V_j$. We have a homeomorphism

$$V_{\text{YM}}^{\ell,0}(SO(2n))_\mu / SO(2n)_{X_\mu} \cong \prod_{j=1}^{r-1}(V_j/U(n_j)) \times V_r/SO(2n_r)$$

and a homotopy equivalence

$$V_{\text{YM}}^{\ell,0}(SO(2n))_\mu{}^{hSO(2n)_{X_\mu}} \sim \prod_{j=1}^{r-1} V_j{}^{hU(n_j)} \times V_r{}^{hSO(2n_r)}.$$

We can decide the topological type of the underlying $SO(2n)$ as in Section 5.1. Then Case 1, Case 2, Case 3 and Case 4 give exactly the same Atiyah-Bott points as in Section 3.4.3.

To simplify the notation, we write

$$\mu = (\mu_1, \ldots, \mu_n) = \Big(\underbrace{\frac{k_1}{n_1}, \ldots, \frac{k_1}{n_1}}_{n_1}, \ldots, \underbrace{\frac{k_r}{n_r}, \ldots, \frac{k_r}{n_r}}_{n_r}\Big)$$

instead of

$$\sqrt{-1}\text{diag}\Big(\frac{k_1}{n_1}J_{n_1}, \ldots, \frac{k_r}{n_r}J_{n_r}\Big),$$

and write

$$\mu = (\mu_1, \ldots, \mu_n) = \Big(\underbrace{\frac{k_1}{n_1}, \ldots, \frac{k_1}{n_1}}_{n_1}, \ldots, \underbrace{\frac{k_{r-1}}{n_{r-1}}, \ldots, \frac{k_{r-1}}{n_{r-1}}}_{n_{r-1}}, \underbrace{\frac{k_r}{n_r}, \ldots, \frac{k_r}{n_r}}_{n_r-1}, \frac{-k_r}{n_r}\Big)$$

instead of

$$\sqrt{-1}\text{diag}\Big(\frac{k_1}{n_1}J_{n_1}, \ldots, \frac{k_{r-1}}{n_{r-1}}J_{n_{r-1}}, \frac{k_r}{n_r}J_{n_r-1}, -\frac{k_r}{n_r}J\Big).$$

Let

$$I_{SO(2n)}^{\pm 1} = \Big\{\mu = \Big(\underbrace{\frac{k_1}{n_1}, \ldots, \frac{k_1}{n_1}}_{n_1}, \ldots, \underbrace{\frac{k_{r-1}}{n_{r-1}}, \ldots, \frac{k_{r-1}}{n_{r-1}}}_{n_{r-1}}, k_r\Big) \Big| \; n_j \in \mathbb{Z}_{>0}, \; k_j \in \mathbb{Z}$$

$$n_1 + \cdots + n_{r-1} + 1 = n, \; \frac{k_1}{n_1} > \cdots > \frac{k_{r-1}}{n_{r-1}} > |k_r| \geq 0, \; (-1)^{k_1 + \cdots + k_r} = \pm 1\Big\}$$

$$\bigcup \Big\{\mu = \Big(\underbrace{\frac{k_1}{n_1}, \ldots, \frac{k_1}{n_1}}_{n_1}, \ldots, \underbrace{\frac{k_{r-1}}{n_{r-1}}, \ldots, \frac{k_{r-1}}{n_{r-1}}}_{n_{r-1}}, \underbrace{\frac{k_r}{n_r}, \ldots, \frac{k_r}{n_r}}_{n_r-1}, \pm\frac{k_r}{n_r}\Big) \Big| \; n_j \in \mathbb{Z}_{>0},$$

$$k_j \in \mathbb{Z}, n_r \in \mathbb{Z}_{>1}, n_1 + \cdots + n_r = n, \frac{k_1}{n_1} > \cdots > \frac{k_r}{n_r} > 0, (-1)^{k_1 + \cdots + k_r} = \pm 1\Big\}$$

$$I^0_{SO(2n)} = \Big\{\mu = \Big(\underbrace{\frac{k_1}{n_1},\ldots,\frac{k_1}{n_1}}_{n_1},\ldots,\underbrace{\frac{k_{r-1}}{n_{r-1}},\ldots,\frac{k_{r-1}}{n_{r-1}}}_{n_{r-1}},\underbrace{0,\ldots,0}_{n_r}\Big)\Big|\ n_j\in\mathbb{Z}_{>0},$$

$$n_r\in\mathbb{Z}_{>1},\ n_1+\cdots+n_r=n,\ k_j\in\mathbb{Z},\ \frac{k_1}{n_1}>\cdots>\frac{k_{r-1}}{n_{r-1}}>0\Big\}.$$

From the above discussion, we conclude that

PROPOSITION 6.1. *Suppose that $\ell\geq 1$.*

(i) *If* $\mu = \Big(\underbrace{\frac{k_1}{n_1},\ldots,\frac{k_1}{n_1}}_{n_1},\ldots,\underbrace{\frac{k_{r-1}}{n_{r-1}},\ldots,\frac{k_{r-1}}{n_{r-1}}}_{n_{r-1}},k_r\Big)\in I^{\pm 1}_{SO(2n)}$, *or*

$$\mu = \Big(\underbrace{\frac{k_1}{n_1},\ldots,\frac{k_1}{n_1}}_{n_1},\ldots,\underbrace{\frac{k_{r-1}}{n_{r-1}},\ldots,\frac{k_{r-1}}{n_{r-1}}}_{n_{r-1}},\underbrace{\frac{k_r}{n_r},\ldots,\frac{k_r}{n_r}}_{n_r-1},\pm\frac{k_r}{n_r}\Big)\in I^{\pm 1}_{SO(2n)},$$

then $X^{\ell,0}_{\mathrm{YM}}(SO(2n))_\mu = X^{\ell,0}_{\mathrm{YM}}(SO(2n+1))^{\pm 1}_\mu$ is nonempty and connected. We have a homeomorphism

$$X^{\ell,0}_{\mathrm{YM}}(SO(2n))_\mu/SO(2n)\cong\prod_{j=1}^{r}\mathcal{M}(\Sigma^\ell_0,P^{n_j,-k_j})$$

and a homotopy equivalence

$$X^{\ell,0}_{\mathrm{YM}}(SO(2n))_\mu{}^{hSO(2n)}\sim\prod_{j=1}^{r}\Big(X^{\ell,0}_{\mathrm{YM}}(U(n_j))_{-\frac{k_j}{n_j},\ldots,-\frac{k_j}{n_j}}\Big)^{hU(n_j)}.$$

(ii) *If* $\mu = \Big(\underbrace{\frac{k_1}{n_1},\ldots,\frac{k_1}{n_1}}_{n_1},\ldots,\underbrace{\frac{k_{r-1}}{n_{r-1}},\ldots,\frac{k_{r-1}}{n_{r-1}}}_{n_{r-1}},\underbrace{0,\ldots,0}_{n_r}\Big)\in I^0_{SO(2n)}$,

then $X^{\ell,0}_{\mathrm{YM}}(SO(2n))_\mu$ has two connected components (from both bundles over Σ^ℓ_0)

$$X^{\ell,0}_{\mathrm{YM}}(SO(2n))^{+1}_\mu\quad\text{and}\quad X^{\ell,0}_{\mathrm{YM}}(SO(2n))^{-1}_\mu.$$

We have a homeomorphism

$$X^{\ell,0}_{\mathrm{YM}}(SO(2n))^{\pm 1}_\mu/SO(2n)\cong\prod_{j=1}^{r-1}\mathcal{M}(\Sigma^\ell_0,P^{n_j,-k_j})\times\mathcal{M}\Big(\Sigma^\ell_0,P^{\pm(-1)^{k_1+\cdots+k_{r-1}}}_{SO(2n_r)}\Big)$$

and a homotopy equivalence

$$X^{\ell,0}_{\mathrm{YM}}(SO(2n))^{\pm 1}_\mu{}^{hSO(2n)}\sim\prod_{j=1}^{r-1}\Big(X^{\ell,0}_{\mathrm{YM}}(U(n_j))_{-\frac{k_j}{n_j},\ldots,-\frac{k_j}{n_j}}\Big)^{hU(n_j)}\times$$
$$\Big(X_{\mathrm{flat}}(SO(2n_r))^{\pm(-1)^{k_1+\cdots+k_{r-1}}}\Big)^{hSO(2n_r)}.$$

PROPOSITION 6.2. *Suppose that $\ell\geq 1$. The connected components of the representation variety $X^{\ell,0}_{\mathrm{YM}}(SO(2n))^{\pm 1}$ are*

$$\{X^{\ell,0}_{\mathrm{YM}}(SO(2n))_\mu\mid\mu\in I^{\pm 1}_{SO(2n)}\}\cup\{X^{\ell,0}_{\mathrm{YM}}(SO(2n))^{\pm 1}_\mu\mid\mu\in I^0_{SO(2n)}\}.$$

The following is an immediate consequence of Proposition 6.1.

THEOREM 6.3. *Suppose that* $\ell \geq 1$.

(i) *If* $\mu = \Big(\underbrace{\frac{k_1}{n_1},\dots,\frac{k_1}{n_1}}_{n_1},\dots,\underbrace{\frac{k_{r-1}}{n_{r-1}},\dots,\frac{k_{r-1}}{n_{r-1}}}_{n_{r-1}},k_r\Big) \in I^{\pm 1}_{SO(2n)}$, *or*

$$\mu = \Big(\underbrace{\frac{k_1}{n_1},\dots,\frac{k_1}{n_1}}_{n_1},\dots,\underbrace{\frac{k_{r-1}}{n_{r-1}},\dots,\frac{k_{r-1}}{n_{r-1}}}_{n_{r-1}},\underbrace{\frac{k_r}{n_r},\dots,\frac{k_r}{n_r}}_{n_r-1},\pm\frac{k_r}{n_r}\Big) \in I^{\pm 1}_{SO(2n)},\ \textit{then}$$

$$P_t^{SO(2n)}\Big(X^{\ell,0}_{\mathrm{YM}}(SO(2n))_\mu\Big) = \prod_{j=1}^{r} P_t^{U(n_j)}\Big(X^{\ell,0}_{\mathrm{YM}}(U(n_i))_{-\frac{k_j}{n_j},\dots,-\frac{k_j}{n_j}}\Big).$$

(ii) *If* $\mu = \Big(\underbrace{\frac{k_1}{n_1},\dots,\frac{k_1}{n_1}}_{n_1},\dots,\underbrace{\frac{k_{r-1}}{n_{r-1}},\dots,\frac{k_{r-1}}{n_{r-1}}}_{n_{r-1}},\underbrace{0,\dots,0}_{n_r}\Big) \in I^{0}_{SO(2n)}$, *then*

$$\begin{aligned} &P_t^{SO(2n)}\Big(X^{\ell,0}_{\mathrm{YM}}(SO(2n))^{\pm 1}_\mu\Big)\\ =\ &\prod_{j=1}^{r-1} P_t^{U(n_j)}\Big(X^{\ell,0}_{\mathrm{YM}}(U(n_j))_{-\frac{k_j}{n_j},\dots,-\frac{k_j}{n_j}}\Big)\cdot P_t^{SO(2n_r)}\Big(X^{\ell,0}_{\mathrm{flat}}(SO(2n_r))^{\pm(-1)^{k_1+\cdots+k_{r-1}}}\Big). \end{aligned}$$

6.2. Equivariant Poincaré series

Recall from Section 3.4.3:

$$\begin{aligned} &\Delta = \{\alpha_i = \theta_i - \theta_{i+1} \mid i = 1,\dots,n-1\} \cup \{\alpha_n = \theta_{n-1}+\theta_n\}\\ &\Delta^\vee = \{\alpha_i^\vee = e_i - e_{i+1} \mid i = 1,\dots,n-1\} \cup \{\alpha_n^\vee = e_{n-1}+e_n\}\\ &\pi_1(H) = \bigoplus_{i=1}^{n}\mathbb{Z}e_i,\quad \Lambda = \bigoplus_{i=1}^{n-1}\mathbb{Z}(e_i - e_{i+1}) \oplus \mathbb{Z}(e_{n-1}+e_n),\\ &\pi_1(SO(2n)) = \langle e_n\rangle \cong \mathbb{Z}/2\mathbb{Z} \end{aligned}$$

We now apply Theorem 4.4 to the case $G_{\mathbb{R}} = SO(2n)$.

$$\begin{aligned} &\varpi_{\alpha_i} = \theta_1 + \cdots + \theta_i,\quad i = 1,\dots,n-2\\ &\varpi_{\alpha_{n-1}} = \frac{1}{2}(\theta_1 + \cdots + \theta_{n-1} - \theta_n),\quad \varpi_{\alpha_n} = \frac{1}{2}(\theta_1 + \cdots + \theta_{n-1} + \theta_n)\\ &\varpi_{\alpha_i}(ke_n) = \begin{cases} 0 & i \leq n-2\\ -k/2 & i = n-1\\ k/2 & i = n \end{cases} \end{aligned}$$

We have the following four cases:

Case 1. $\alpha_{n-1}, \alpha_n \in I$: $n_r = 1$

$$
\begin{aligned}
&I = \{\alpha_{n_1}, \alpha_{n_1+n_2}, \dots, \alpha_{n_1+\cdots+n_{r-2}}, \alpha_{n-1}, \alpha_n\}\\
&L^I = GL(n_1,\mathbb{C}) \times \cdots \times GL(n_{r-1},\mathbb{C}) \times GL(1,\mathbb{C}), \quad n_1 + \cdots + n_{r-1} + 1 = n\\
&\dim_{\mathbb{C}} \mathfrak{z}_{L^I} - \dim_{\mathbb{C}} \mathfrak{z}_{SO(2n,\mathbb{C})} = r, \quad \dim_{\mathbb{C}} U^I = \sum_{1\le i<j\le r} n_i n_j + \frac{n(n-1)}{2}\\
&\rho^I = \frac{1}{2}\sum_{i=1}^{r}\Big(n - 2\sum_{j=1}^{i} n_j + n_i\Big)\Big(\sum_{j=1}^{n_i}\theta_{n_1+\cdots+n_{i-1}+j}\Big) + \frac{n-1}{2}(\theta_1 + \cdots + \theta_n)\\
&\langle \rho^I, \alpha^\vee_{n_1+\cdots+n_i}\rangle = \frac{n_i + n_{i+1}}{2}, \quad \text{for } i = 1, \dots, r-2,\\
&\langle \rho^I, \alpha^\vee_{n-1}\rangle = \langle \rho^I, \alpha^\vee_n\rangle = \frac{n_{r-1}+1}{2}
\end{aligned}
$$

Case 2. $\alpha_{n-1} \in I$, $\alpha_n \notin I$: $n_r > 1$

$$
\begin{aligned}
&I = \{\alpha_{n_1}, \alpha_{n_1+n_2}, \dots, \alpha_{n_1+\cdots+n_{r-1}}, \alpha_{n-1}\}\\
&L^I = GL(n_1,\mathbb{C}) \times \cdots \times GL(n_r,\mathbb{C}), \quad n_1 + \cdots + n_r = n\\
&\dim_{\mathbb{C}} \mathfrak{z}_{L^I} - \dim_{\mathbb{C}} \mathfrak{z}_{SO(2n,\mathbb{C})} = r, \quad \dim_{\mathbb{C}} U^I = \sum_{1\le i<j\le r} n_i n_j + \frac{n(n-1)}{2}\\
&\rho^I = \frac{1}{2}\sum_{i=1}^{r}\Big(n - 2\sum_{j=1}^{i} n_j + n_i\Big)\Big(\sum_{j=1}^{n_i}\theta_{n_1+\cdots+n_{i-1}+j}\Big) + \frac{n-1}{2}\Big(\sum_{j=1}^{n}\theta_j\Big) - (n_r-1)\theta_n
\end{aligned}
$$

$$
\langle \rho^I, \alpha^\vee_{n_1+\cdots+n_i}\rangle = \frac{n_i + n_{i+1}}{2} \text{ for } i = 1, \dots, r-1, \quad \langle \rho^I, \alpha^\vee_{n-1}\rangle = n_r - 1
$$

Case 3. $\alpha_{n-1} \notin I$, $\alpha_n \in I$: $n_r > 1$

$$
\begin{aligned}
&I = \{\alpha_{n_1}, \alpha_{n_1+n_2}, \dots, \alpha_{n_1+n_2+\cdots+n_{r-1}}, \alpha_n\}\\
&L^I = GL(n_1,\mathbb{C}) \times \cdots \times GL(n_r,\mathbb{C}), \quad n_1 + \cdots + n_r = n\\
&\dim_{\mathbb{C}} \mathfrak{z}_{L^I} - \dim_{\mathbb{C}} \mathfrak{z}_{SO(2n,\mathbb{C})} = r, \quad \dim_{\mathbb{C}} U^I = \sum_{1\le i<j\le r} n_i n_j + \frac{n(n-1)}{2}\\
&\rho^I = \frac{1}{2}\sum_{i=1}^{r}\Big(n - 2\sum_{j=1}^{i} n_j + n_i\Big)\Big(\sum_{j=1}^{n_i}\theta_{n_1+\cdots+n_{i-1}+j}\Big) + \frac{n-1}{2}(\theta_1 + \cdots + \theta_n)
\end{aligned}
$$

$$
\langle \rho^I, \alpha^\vee_{n_1+\cdots+n_i}\rangle = \frac{n_i + n_{i+1}}{2} \text{ for } i = 1, \dots, r-1, \quad \langle \rho^I, \alpha^\vee_n\rangle = n_r - 1
$$

Case 4. $\alpha_{n-1} \notin I$, $\alpha_n \notin I$: $n_r > 1$

$$I = \{\alpha_{n_1}, \alpha_{n_1+n_2}, \dots, \alpha_{n_1+n_2+\cdots+n_{r-1}}\}$$
$$L^I = GL(n_1,\mathbb{C}) \times \cdots \times GL(n_{r-1},\mathbb{C}) \times SO(2n_r), \quad n_1 + \cdots + n_r = n$$
$$\dim_{\mathbb{C}} \mathfrak{z}_{L^I} - \dim_{\mathbb{C}} \mathfrak{z}_{SO(2n,\mathbb{C})} = r - 1,$$
$$\dim_{\mathbb{C}} U^I = \sum_{1\le i<j\le r} n_i n_j + \frac{n(n-1) - n_r(n_r-1)}{2}$$
$$\rho^I = \frac{1}{2}\sum_{i=1}^{r}\left(n - 2\sum_{j=1}^{i} n_j + n_i\right)\left(\sum_{j=1}^{n_i}\theta_{n_1+\cdots+n_{i-1}+j}\right)$$
$$+\frac{n-1}{2}(\theta_1 + \cdots + \theta_{n_1+\cdots+n_{r-1}}) + \frac{n-n_r}{2}(\theta_{n_1+\cdots+n_{r-1}+1} + \cdots + \theta_n)$$
$$\langle \rho^I, \alpha^\vee_{n_1+\cdots+n_i}\rangle = \frac{n_i + n_{i+1}}{2}, \quad \text{for } i = 1, \dots, r-2,$$
$$\langle \rho^I, \alpha^\vee_{n_1+\cdots+n_{r-1}}\rangle = \frac{n_{r-1} + 2n_r - 1}{2}$$

We have the following closed formula for the $SO(2n)$-equivariant Poincaré series of the representation variety of flat $SO(2n)$-connections over Σ_0^ℓ:

THEOREM 6.4. $n \ge 2$

$$P_t^{SO(2n)}(X^{\ell,0}_{\text{flat}}(SO(2n))^{(-1)^k}) =$$
$$\sum_{r=2}^{n} \sum_{\substack{n_1,\dots,n_r \in \mathbb{Z}_{>0} \\ \sum n_j = n,\, n_r = 1}} (-1)^r \prod_{i=1}^{r} \frac{\prod_{j=1}^{n_i}(1+t^{2j-1})^{2\ell}}{(1-t^{2n_i})\prod_{j=1}^{n_i-1}(1-t^{2j})^2}$$
$$\cdot \frac{t^{(\ell-1)(2\sum_{i<j} n_i n_j + n(n-1))}}{\left[\prod_{i=1}^{r-1}(1-t^{2(n_i+n_{i+1})})\right](1-t^{2(n_{r-1}+1)})} t^{2\sum_{i=1}^{r-2}(n_i+n_{i+1}) + 4(n_{r-1}+1)\langle k/2\rangle}$$
$$+\sum_{r=1}^{n-1} \sum_{\substack{n_1,\dots,n_r \in \mathbb{Z}_{>0} \\ \sum n_j = n,\, n_r > 1}} \left(2(-1)^r \prod_{i=1}^{r} \frac{\prod_{j=1}^{n_i}(1+t^{2j-1})^{2\ell}}{(1-t^{2n_i})\prod_{j=1}^{n_i-1}(1-t^{2j})^2} \right.$$
$$\cdot \frac{t^{(\ell-1)(2\sum_{i<j} n_i n_j + n(n-1))}}{\left[\prod_{i=1}^{r-1}(1-t^{2(n_i+n_{i+1})})\right](1-t^{4(n_r-1)})} \cdot t^{2\sum_{i=1}^{r-1}(n_i+n_{i+1}) + 4(n_r-1)\langle k/2\rangle}$$
$$+(-1)^{r-1} \prod_{i=1}^{r-1} \frac{\prod_{j=1}^{n_i}(1+t^{2j-1})^{2\ell}}{(1-t^{2n_i})\prod_{j=1}^{n_i-1}(1-t^{2j})^2} \cdot \frac{(1+t^{2n_r-1})^{2\ell}\prod_{j=1}^{n_r-1}(1+t^{4j-1})^{2\ell}}{(1-t^{2n_r-2})(1-t^{2n_r})\prod_{j=1}^{2n_r-2}(1-t^{2j})}$$
$$\left. \cdot \frac{t^{(\ell-1)(2\sum_{i<j} n_i n_j + n(n-1) - n_r(n_r-1))}}{\left[\prod_{i=1}^{r-2}(1-t^{2(n_i+n_{i+1})})\right](1-\epsilon(r)t^{2(n_{r-1}+2n_r-1)})} t^{2\sum_{i=1}^{r-2}(n_i+n_{i+1}) + 2\epsilon(r)(n_{r-1}+2n_r-1)} \right)$$

where

$$\epsilon(r) = \begin{cases} 0 & r = 1 \\ 1 & r > 1 \end{cases}$$

REMARK 6.5. For $n \ge 2$, we have

$$P_t^{SO(2n)}(X^{\ell,0}_{\text{flat}}(SO(2n))^{+1}) = P_t^{Spin(2n)}(X^{\ell,0}_{\text{flat}}(Spin(2n))),$$

so Theorem 6.4 also gives a formula for $X^{\ell,0}_{\text{flat}}(Spin(2n))$.

EXAMPLE 6.6.

$$
\begin{aligned}
&P_t^{SO(4)}(X^{\ell,0}_{\text{flat}}(SO(4))^{+1}) = P_t^{Spin(4)}(X^{\ell,0}_{\text{flat}}(Spin(4)))\\
=\ & \frac{(1+t)^{4\ell}t^{4\ell+4}}{(1-t^2)^2(1-t^4)^2} - 2\frac{(1+t)^{2\ell}(1+t^3)^{2\ell}t^{2\ell+2}}{(1-t^2)^2(1-t^4)^2} + \frac{(1+t^3)^{4\ell}}{(1-t^2)^2(1-t^4)^2}\\
=\ & \frac{1}{(1-t^2)^2(1-t^4)^2}\left((1+t^3)^{4\ell} - 2t^{2\ell+2}(1+t)^{2\ell}(1+t^3)^{2\ell} + t^{4\ell+4}(1+t)^{4\ell}\right)\\
&P_t^{SO(4)}(X^{\ell,0}_{\text{flat}}(SO(4))^{-1})\\
=\ & \frac{(1+t)^{4\ell}t^{4\ell+2}}{(1-t^2)^2(1-t^4)^2} - 2\frac{(1+t)^{2\ell}(1+t^3)^{2\ell}t^{2\ell}}{(1-t^2)^2(1-t^4)^2} + \frac{(1+t^3)^{4\ell}}{(1-t^2)^2(1-t^4)^2}\\
=\ & \frac{(1+t)^{2\ell}}{(1-t^2)^2(1-t^4)^2}\left((1+t^3)^{4\ell} - 2t^{2\ell}(1+t)^{2\ell}(1+t^3)^{2\ell} + t^{4\ell}(1+t)^{4\ell}\right)
\end{aligned}
$$

Note that $Spin(4) = SU(2)\times SU(2)$, so

$$P_t^{Spin(4)}(X^{\ell,0}_{\text{flat}}(Spin(4))) = \left(P_t^{SU(2)}(X^{\ell,0}_{\text{flat}}(SU(2)))\right)^2$$

as expected, where $P_t^{SU(2)}(X^{\ell,0}_{\text{flat}}(SU(2)))$ is calculated in Example 4.7.

EXAMPLE 6.7.

$$
\begin{aligned}
&P_t^{SO(6)}(X^{\ell,0}_{\text{flat}}(SO(6))^{+1}) = P_t^{Spin(6)}(X^{\ell,0}_{\text{flat}}(Spin(6)))\\
=\ & \frac{(1+t)^{4\ell}(1+t^3)^{2\ell}t^{10\ell+2}}{(1-t^2)^3(1-t^4)(1-t^6)^2} - \frac{(1+t)^{6\ell}t^{12\ell}}{(1-t^2)^3(1-t^4)^3}\\
&-2\frac{(1+t)^{2\ell}(1+t^3)^{2\ell}(1+t^5)^{2\ell}t^{6\ell+2}}{(1-t^2)^2(1-t^4)^2(1-t^6)(1-t^8)} + 2\frac{(1+t)^{4\ell}(1+t^3)^{2\ell}t^{10\ell}}{(1-t^2)^3(1-t^4)^2(1-t^6)}\\
&+\frac{(1+t^3)^{2\ell}(1+t^5)^{2\ell}(1+t^7)^{2\ell}}{(1-t^2)(1-t^4)^2(1-t^6)^2(1-t^8)} - \frac{(1+t)^{2\ell}(1+t^3)^{4\ell}t^{8\ell}}{(1-t^2)^3(1-t^4)^2(1-t^8)}\\
&P_t^{SO(6)}(X^{\ell,0}_{\text{flat}}(SO(6))^{-1})\\
=\ & \frac{(1+t)^{4\ell}(1+t^3)^{2\ell}t^{10\ell-4}}{(1-t^2)^3(1-t^4)(1-t^6)^2} - \frac{(1+t)^{6\ell}t^{12\ell-4}}{(1-t^2)^3(1-t^4)^3}\\
&-2\frac{(1+t)^{2\ell}(1+t^3)^{2\ell}(1+t^5)^{2\ell}t^{6\ell-2}}{(1-t^2)^2(1-t^4)^2(1-t^6)(1-t^8)} + 2\frac{(1+t)^{4\ell}(1+t^3)^{2\ell}t^{10\ell-2}}{(1-t^2)^3(1-t^4)^2(1-t^6)}\\
&+\frac{(1+t^3)^{2\ell}(1+t^5)^{2\ell}(1+t^7)^{2\ell}}{(1-t^2)(1-t^4)^2(1-t^6)^2(1-t^8)} - \frac{(1+t)^{2\ell}(1+t^3)^{4\ell}t^{8\ell}}{(1-t^2)^3(1-t^4)^2(1-t^8)}
\end{aligned}
$$

Note that $Spin(6) = SU(4)$, so

$$P_t^{Spin(6)}(X^{\ell,0}_{\text{flat}}(Spin(6))) = P_t^{SU(4)}(X^{\ell,0}_{\text{flat}}(SU(4)))$$

as expected, where $P_t^{SU(4)}(X^{\ell,0}_{\text{flat}}(SU(4)))$ is calculated in Example 4.8.

6.3. $SO(4m+2)$-connections on nonorientable surfaces

In this section, we consider $SO(2n)$ where $n = 2m+1$ is odd, so that

$$\overline{C_0^\tau} = \{\sqrt{-1}\mathrm{diag}(t_1 J,\ldots,t_{2m}J, 0J) \mid t_1 \geq \cdots \geq t_{2m} \geq 0\}.$$

Any $\mu \in \overline{C}_0^{\tau}$ is of the form

$$\mu = \sqrt{-1}\mathrm{diag}(\lambda_1 J_{n_1}, \ldots, \lambda_{r-1} J_{n_{r-1}}, 0 J_{n_r}),$$

where $\lambda_1 > \cdots > \lambda_{r-1} > 0$ and $n_r > 0$. We have

$$SO(2n)_{X_\mu} \cong \Phi(U(n_1)) \times \cdots \times \Phi(U(n_{r-1})) \times SO(2n_r),$$

where $X_\mu = -2\pi\sqrt{-1}\mu$.

Given $\mu \in \overline{C}_0^{\tau}$, let

$$\epsilon_\mu = \mathrm{diag}(H_{n-n_r}, (-1)^{(n-n_r)} I_1, I_{2n_r-1}).$$

Then $\mathrm{Ad}(\epsilon_\mu)X_\mu = -X_\mu$. Note that $n_r \geq 1$.

Suppose that

$$(a_1, b_1, \ldots, a_\ell, b_\ell, \epsilon_\mu c', X_\mu/2) \in X_{\mathrm{YM}}^{\ell,1}(SO(2n)).$$

Then

$$\exp(X_\mu/2)\epsilon_\mu c' \epsilon_\mu c' = \prod_{i=1}^{\ell} [a_i, b_i]$$

where $a_i,\ b_i,\ c' \in \Phi(U(n_1)) \times \cdots \times \Phi(U(n_{r-1})) \times SO(2n_r)$.

Let $L : \mathbb{R}^{2(n-n_r)} \to \mathbb{C}^{n-n_r}$ be defined as in Section 5.1, and let

$$\begin{aligned} X'_\mu &= L \circ \big(2\pi \mathrm{diag}(\lambda_1 J_{n_1}, \ldots, \lambda_{r-1} J_{n_{r-1}})\big) \circ L^{-1} \\ &= 2\pi\sqrt{-1}\mathrm{diag}(\lambda_1 I_{n_1}, \cdots, \lambda_{r-1} I_{n_{r-1}}) \in \mathfrak{u}(n_1) \times \cdots \times \mathfrak{u}(n_{r-1}). \end{aligned}$$

Then the condition on X'_μ is

$$\exp(X'_\mu/2)\bar{c'} c' = \prod_{i=1}^{\ell} [a_i, b_i] \in SU(n_1) \times \cdots \times SU(n_{r-1})$$

where $a_i,\ b_i,\ c' \in U(n_1) \times \cdots \times U(n_{r-1})$, and $\bar{c'}$ is the complex conjugate of c'. In order that this is nonempty, we need $1 = \det(e^{\pi\sqrt{-1}\lambda_j} I_{n_j})$, i.e.,

$$\lambda_j = \frac{2k_j}{n_j}, \quad k_j \in \mathbb{Z}, \quad j = 1, \ldots, r-1. \tag{6.4}$$

Similarly, suppose that $(a_1, b_1, \ldots, a_\ell, b_\ell, d, \epsilon_\mu c', X_\mu/2) \in X_{\mathrm{YM}}^{\ell,2}(SO(2n)$. Then

$$\exp(X_\mu/2)(\epsilon_\mu c') d (\epsilon_\mu c')^{-1} d = \prod_{i=1}^{\ell} [a_i, b_i]$$

where $a_i,\ b_i,\ d,\ c' \in \Phi(U(n_1)) \times \cdots \times \Phi(U(n_{r-1})) \times SO(2n_r)$. The condition on X'_μ is

$$\exp(X'_\mu/2)\bar{c'} \bar{d} \bar{c'}^{-1} d \in SU(n_1) \times \cdots \times SU(n_{r-1}).$$

Again, we need

$$\lambda_j = \frac{2k_j}{n_j}, \quad k_j \in \mathbb{Z}, \quad j = 1, \ldots, r-1.$$

We conclude that for nonorientable surfaces,

$$\mu = \sqrt{-1}\mathrm{diag}\Big(\frac{2k_1}{n_1} J_{n_1}, \ldots, \frac{2k_{r-1}}{n_{r-1}} J_{n_{r-1}}, 0 J_{n_r}\Big), \text{where } k_j \in \mathbb{Z},\ \frac{k_i}{n_i} > \frac{k_{i+1}}{n_{i+1}} > 0, n_r > 0.$$

For each μ, the ϵ_μ-reduced representation varieties are

$$\begin{aligned} V^{\ell,1}_{\mathrm{YM}}(SO(2n))_\mu &= \{(a_1,b_1,\ldots,a_\ell,b_\ell,c') \in SO(2n)^{2\ell+1}_{X_\mu} \mid \\ &\quad \prod_{i=1}^{\ell}[a_i,b_i] = \exp(\frac{X_\mu}{2})\epsilon_\mu c'\epsilon_\mu c'\}, \\ V^{\ell,2}_{\mathrm{YM}}(SO(2n))_\mu &= \{(a_1,b_1,\ldots,a_\ell,b_\ell,d,c') \in SO(2n)^{2\ell+2}_{X_\mu} \mid \\ &\quad \prod_{i=1}^{\ell}[a_i,b_i] = \exp(\frac{X_\mu}{2})\epsilon_\mu c' d(\epsilon_\mu c')^{-1}d\}. \end{aligned}$$

For $i = 1,\cdots,\ell$, write

$$\begin{aligned} a_i &= \mathrm{diag}(A^i_1,\ldots,A^i_r), \quad b_i = \mathrm{diag}(B^i_1,\ldots,B^i_r), \\ c' &= \mathrm{diag}(C_1,\ldots,C_r), \quad d = \mathrm{diag}(D_1,\ldots,D_r), \end{aligned}$$

where A^i_j, B^i_j, C_j, $D_j \in \Phi(U(n_j))$ for $j = 1,\ldots,r-1$, and A^i_r, B^i_r, C_r, $D_r \in SO(2n_r)$.

$i = 1$. Define V_j as in (5.7), and define

$$V_r = \Big\{(A^1_r,B^1_r,\ldots,A^\ell_r,B^\ell_r,C_r) \in SO(2n_r)^{2\ell+1} \mid \prod_{i=1}^{\ell}[A^i_r,B^i_r] = (\epsilon C_r)^2\Big\}, \tag{6.5}$$

where $\epsilon = \mathrm{diag}((-1)^{n-n_r}, I_{2n_r-1})$, $\det(\epsilon) = (-1)^{n-n_r}$. Let $C'_r = \epsilon C_r$. We see that

$$\begin{aligned} V_r &\cong \Big\{(A^1_r,B^1_r,\ldots,A^\ell_r,B^\ell_r,C'_r) \in SO(2n_r)^{2\ell} \times O(2n_r) \mid \\ &\quad \prod_{i=1}^{\ell}[A^i_r,B^i_r] = (C'_r)^2, \det(C'_r) = (-1)^{n-n_r}\Big\} \\ &\cong V^{\ell,1}_{O(2n_r),(-1)^{n-n_r}} \end{aligned}$$

where $V^{\ell,1}_{O(n),\pm 1}$ is the twisted representation variety defined in (4.11) of Section 4.7. $V^{\ell,1}_{O(n),\pm 1}$ is nonempty if $\ell \geq 2$. We have shown that $V^{\ell,1}_{O(n),\pm 1}$ is disconnected with two components $V^{\ell,1,+1}_{O(n),\pm 1}$ and $V^{\ell,1,-1}_{O(n),\pm 1}$ if $\ell \geq 2$ and $n > 2$ (Proposition 4.14).

We have

$$V^{\ell,1}_{\mathrm{YM}}(SO(2n))_\mu = \prod_{j=1}^{r} V_j.$$

We define a $U(n_j)$-action on $V_j = \tilde{V}^{\ell,1}_{n_j,-k_j}$ by (4.9) of Section 4.6, and an $SO(2n_r)$-action on $V_r = V^{\ell,1}_{O(2n_r),(-1)^{n-n_r}}$ by (4.13) of Section 4.7. Then we have a homeomorphism

$$V^{\ell,1}_{\mathrm{YM}}(SO(2n))_\mu/SO(2n)_{X_\mu} \cong \prod_{j=1}^{r-1}(V_j/U(n_j)) \times V_r/SO(2n_r)$$

and a homotopy equivalence

$$V^{\ell,1}_{\mathrm{YM}}(SO(2n))_\mu^{hSO(2n)_{X_\mu}} \sim \prod_{j=1}^{r-1} V_j^{hU(n_j)} \times V_r^{hSO(2n_r)}.$$

$i = 2$. Define V_j as in (5.9), and define

(6.6)

$$V_r = \Big\{(A_r^1, B_r^1, \ldots, A_r^\ell, B_r^\ell, D_r, C_r) \in SO(2n_r)^{2\ell+2} \mid \prod_{i=1}^{\ell}[A_r^i, B_r^i] = \epsilon C_r D_r (\epsilon C_r)^{-1} D_r\Big\}$$

where $\epsilon = \mathrm{diag}((-1)^{n-n_r} I_1, I_{2n_r-1})$, $\det(\epsilon) = (-1)^{n-n_r}$. Let $C'_r = \epsilon C_r$. We see that

$$\begin{aligned} V_r &\cong \Big\{(A_r^1, B_r^1, \ldots, A_r^\ell, B_r^\ell, D_r, C'_r) \in SO(2n_r)^{2\ell+1} \times O(2n_r) \mid \\ &\quad \prod_{i=1}^{\ell}[A_r^i, B_r^i] = C'_r D_r C_r'^{-1} D_r, \det(C'_r) = (-1)^{n-n_r}\Big\} \\ &\cong V^{\ell,2}_{O(2n_r),(-1)^{n-n_r}} \end{aligned}$$

where $V^{\ell,2}_{O(n),\pm1}$ is the twisted representation variety defined in (4.12) of Section 4.7. $V^{\ell,2}_{O(n),\pm1}$ is nonempty if $\ell \geq 4$. We have shown that $V^{\ell,2}_{O(n),\pm1}$ is disconnected with two components $V^{\ell,1,+1}_{O(n),\pm1}$ and $V^{\ell,1,-1}_{O(n),\pm1}$ if $\ell \geq 4$ and $n > 2$ (Proposition 4.14).

We have

$$V^{\ell,2}_{\mathrm{YM}}(SO(2n))_\mu = \prod_{j=1}^{r} V_j.$$

We define a $U(n_j)$-action on $V_j = \tilde{V}^{\ell,2}_{n_j,-k_j}$ by (4.10) of Section 4.6, and an $SO(2n_r)$-action on $V_r = V^{\ell,2}_{O(2n_r),(-1)^{n-n_r}}$ by (4.14) of Section 4.7. Then we have a homeomorphism

$$V^{\ell,2}_{\mathrm{YM}}(SO(2n))_\mu / SO(2n)_{X_\mu} \cong \prod_{j=1}^{r-1}(V_j/U(n_j)) \times V_r/SO(2n_r)$$

and a homotopy equivalence

$$V^{\ell,2}_{\mathrm{YM}}(SO(2n))_\mu{}^{hSO(2n)_{X_\mu}} \sim \prod_{j=1}^{r-1} V_j{}^{hU(n_j)} \times V_r{}^{hSO(2n_r)}.$$

Note that, $V^{\ell,i}_{O(2),-1} \cong \tilde{V}^{\ell,i}_{1,0} \cong U(1)^{2\ell+i}$ is connected as mentioned in Section 4.7.

We have seen that $V^{\ell,i}_{\mathrm{YM}}(SO(2n))_\mu$ is disconnected with two connected components if $\ell \geq 2i$ and $n_r \geq 1$ (notice that when $n_r = 1$, $n - n_r = 2m$ is even). To determine the underlying topological $SO(2n)$-bundle P for each component, we consider four special cases.

Case 1. Assuming that $n_r > 1$, we consider special points

$$(a_1, b_1, \ldots, a_\ell, b_\ell, c) \in V^{\ell,1}_{\mathrm{YM}}(SO(2n))_\mu, \quad (a_1, b_1, \ldots, a_\ell, b_\ell, d, c) \in V^{\ell,2}_{\mathrm{YM}}(SO(2n))_\mu,$$

where

$$\begin{aligned} &a_i = \mathrm{diag}(A_1^i, \ldots, A_{r-1}^i, I_{2n_r}), \quad b_i = \mathrm{diag}(B_1^i, \ldots, B_{r-1}^i, I_{2n_r}), \\ &c = \epsilon_\mu = \mathrm{diag}(H_{n-n_r}, (-1)^{n-n_r} I_1, I_{2n_r-1}), \quad d = I_{2n}. \end{aligned}$$

Let $\epsilon_1 = \mathrm{diag}((-1)^{n-n_r} I_1, I_{2n_r-1})$. Then

$$(A^i_j, B^i_j, \ldots, A^i_j, B^i_j) \in X^{\ell,0}_{\mathrm{YM}}(U(n_j))_{-\frac{k_j}{n_j},\ldots,-\frac{k_j}{n_j}}, \quad j = 1, \ldots, r-1,$$

$$(I_{2n_r}, \ldots, I_{2n_r}, \epsilon_1) \in V^{\ell,1,(-1)^{n-n_r}}_{O(2n_r),(-1)^{n-n_r}}, (I_{2n_r}, \ldots, I_{2n_r}, I_{2n_r}, \epsilon_1) \in V^{\ell,2,1}_{O(2n_r),(-1)^{n-n_r}}.$$

We have $P = P_1 \times P_2$, where P_1 is an $SO(2(n-n_r)+1)$-bundle, and P_2 is an $SO(2n_r - 1)$-bundle with trivial holonomies I_{2n_r-1}. We have

$$w_2(P) = w_2(P_1) = k_1 + \cdots + k_{r-1} + i\frac{(n-n_r)(n-n_r+1)}{2} \pmod 2,$$

where the second equality follows from the argument in Section 5.3.

Case 2. Assuming that $n_r > 1$, as in *Case 1*, we consider special points

$$(a_1, b_1, \ldots, a_\ell, b_\ell, c) \in V^{\ell,1}_{\mathrm{YM}}(SO(2n))_\mu, \quad (a_1, b_1, \ldots, a_\ell, b_\ell, d, c) \in V^{\ell,2}_{\mathrm{YM}}(SO(2n))$$

where

$$a_i = \mathrm{diag}(A^i_1, \ldots, A^i_{r-1}, I_{2n_r}), \quad b_i = \mathrm{diag}(B^i_1, \ldots, B^i_{r-1}, I_{2n_r}),$$
$$c = \mathrm{diag}(H_{n-n_r}, (-1)^{(n-n_r)} I_1, -I_2, I_{2n_r-3}), \quad d = \mathrm{diag}(I_{2(n-n_r)+1}, -I_2, I_{2n_r-3}).$$

Let $\epsilon_1 = \mathrm{diag}((-1)^{n-n_r} I_1, -I_2, I_{2n_r-3})$, $\epsilon_2 = \mathrm{diag}(I_1, -I_2, I_{2n_r-3})$, and $\epsilon = \mathrm{diag}(-I_2, I_{2n_r-3})$. Then

$$(A^i_j, B^i_j, \ldots, A^i_j, B^i_j) \in X^{\ell,0}_{\mathrm{YM}}(U(n_j))_{-\frac{k_j}{n_j},\ldots,-\frac{k_j}{n_j}}, \quad j = 1, \ldots, r-1,$$

$$(I_{2n_r}, \ldots, I_{2n_r}, \epsilon_1) \in V^{\ell,1,-(-1)^{n-n_r}}_{O(2n_r),(-1)^{n-n_r}}, \quad (I_{2n_r}, \ldots, I_{2n_r}, \epsilon_2, \epsilon_1) \in V^{\ell,2,-1}_{O(2n_r),(-1)^{n-n_r}}.$$

We have $P = P_1 \times P_2$, where P_1 is an $SO(2(n-n_r)+1)$-bundle, and P_2 is an $SO(2n_r - 1)$-bundle with holonomies $a_i = b_i = I_{2n_r-1}$, $c = d = \epsilon$. We can choose the lifting of d and c as $\tilde{d} = \tilde{c} = e_1 e_2$ and $\tilde{c}^2 = \tilde{c}\tilde{d}\tilde{c}^{-1}\tilde{d} = -1$. Thus we have

$$w_2(P_1) = k_1 + \cdots + k_{r-1} + i\frac{(n-n_r)(n-n_r+1)}{2} \pmod 2, \quad w_2(P_2) = 1 \pmod 2,$$

so

$$w_2(P) = w_2(P_1) + w_2(P_2) = k_1 + \cdots + k_{r-1} + i\frac{(n-n_r)(n-n_r+1)}{2} + 1 \pmod 2.$$

Case 3. Assuming that $n_r = 1$ so that $n - n_r = 2m$ is even, we consider special points

$$(a_1, b_1, \ldots, a_\ell, b_\ell, c) \in V^{\ell,1}_{\mathrm{YM}}(SO(2n))_\mu, \quad (a_1, b_1, \ldots, a_\ell, b_\ell, d, c) \in V^{\ell,2}_{\mathrm{YM}}(SO(2n))_\mu,$$

where

$$a_i = \mathrm{diag}(A^i_1, \ldots, A^i_{r-1}, I_2), \quad b_i = \mathrm{diag}(B^i_1, \ldots, B^i_{r-1}, I_2),$$
$$c = \mathrm{diag}(H_{2m}, -I_2), \quad d = \mathrm{diag}(I_{4m}, -I_2).$$

Then

$$(A^i_j, B^i_j, \ldots, A^i_j, B^i_j) \in X^{\ell,0}_{\mathrm{YM}}(U(n_j))_{-\frac{k_j}{n_j},\ldots,-\frac{k_j}{n_j}}, \quad j = 1, \ldots, r-1,$$

$$(I_2, \ldots, I_2, -I_2) \in V^{\ell,1,-1}_{O(2),+1}, \quad (I_2, \ldots, I_2, -I_2, -I_2) \in V^{\ell,2,-1}_{O(2),+1}.$$

We have $P = P_1 \times P_2$, where P_1 is an $SO(4m)$-bundle with holonomies $d = I_{4m}$ and $c = H_{2m}$ with lifting $\tilde{c} = e_2 e_4 \cdots e_{4m}$, and P_2 is an $SO(2)$-bundle with holonomies $a_i = b_i = I_2$ and $c = d = -I_2$ with lifting $\tilde{d} = \tilde{c} = e_1 e_2$. Then we have

$$w_2(P_1) = k_1 + \cdots + k_{r-1} + im \pmod 2, \quad w_2(P_2) = 1 \pmod 2,$$

so

$$w_2(P) = k_1 + \cdots + k_{r-1} + im + 1.$$

Case 4. Assuming that $n_r = 1$ as in *Case 3*, we consider special points

$$(a_1, b_1, \ldots, a_\ell, b_\ell, c) \in V^{\ell,1}_{\mathrm{YM}}(SO(2n))_\mu, \quad (a_1, b_1, \ldots, a_\ell, b_\ell, d, c) \in V^{\ell,2}_{\mathrm{YM}}(SO(2n))_\mu,$$

where

$$\begin{aligned} &a_i = \mathrm{diag}(A^i_1, \ldots, A^i_{r-1}, I_2), \quad b_i = \mathrm{diag}(B^i_1, \ldots, B^i_{r-1}, I_2), \\ &c = \mathrm{diag}(H_{2m}, I_2), \quad d = I_{2n}. \end{aligned}$$

Then

$$(A^i_j, B^i_j, \ldots, A^i_j, B^i_j) \in X^{\ell,0}_{\mathrm{YM}}(U(n_j))_{-\frac{k_j}{n_j}, \ldots, -\frac{k_j}{n_j}}, \quad j = 1, \ldots, r-1,$$

$$(I_2, \ldots, I_2, I_2) \in V^{\ell,1,1}_{O(2),+1}, \quad (I_2, \ldots, I_2, I_2, I_2) \in V^{\ell,2,1}_{O(2),+1}.$$

We have $P = P_1 \times P_2$, where P_1 is an $SO(4m)$-bundle with holonomies $d = I_{4m}$ and $c = H_{2m}$ with lifting $\tilde{c} = e_2 e_4 \cdots e_{4m}$, and P_2 is an $SO(2)$-bundle with trivial holonomies I_2. Then we have

$$w_2(P) = w_2(P_1) = k_1 + \cdots + k_{r-1} + im \text{ (mod 2)}.$$

To summarize, when $n = 2m+1$, we have

$$V^{\ell,i}_{\mathrm{YM}}(SO(2n))^{\pm}_\mu = \prod_{j=1}^{r-1} V_j \times V^{\ell,i,\pm(-1)^{k_1+\cdots+k_{r-1}+i\frac{(n-n_r)(n-n_r-1)}{2}}}_{O(2n_r),(-1)^{n-n_r}},$$

where $V^{\ell,i}_{\mathrm{YM}}(SO(2n))^{\pm}_\mu$ is the ϵ_μ-reduced version of $X^{\ell,i}_{\mathrm{YM}}(SO(2n))^{\pm 1}_\mu$. Note that

$$i\frac{(n-n_r)(n-n_r-1)}{2} \equiv i(m + \frac{n_r(n_r-1)}{2}) \text{ (mod 2)}, \quad n - n_r \equiv n_r - 1 \text{ (mod 2)}.$$

To simplify the notation, we write

$$\mu = (\mu_1, \ldots, \mu_{2m}, 0) = \Big(\underbrace{\frac{2k_1}{n_1}, \ldots, \frac{2k_1}{n_1}}_{n_1}, \ldots, \underbrace{\frac{2k_{r-1}}{n_{r-1}}, \ldots, \frac{2k_{r-1}}{n_{r-1}}}_{n_{r-1}}, \underbrace{0, \ldots, 0}_{n_r}\Big)$$

instead of

$$\sqrt{-1}\mathrm{diag}\Big(\frac{2k_1}{n_1} J_{n_1}, \ldots, \frac{2k_{r-1}}{n_{r-1}} J_{n_{r-1}}, 0 J_{n_r}\Big).$$

Let

$$\begin{aligned} \hat{I}_{SO(4m+2)} \quad = \quad & \Big\{\mu = \Big(\underbrace{\frac{2k_1}{n_1}, \ldots, \frac{2k_1}{n_1}}_{n_1}, \ldots, \underbrace{\frac{2k_{r-1}}{n_{r-1}}, \ldots, \frac{2k_{r-1}}{n_{r-1}}}_{n_{r-1}}, \underbrace{0, \ldots, 0}_{n_r}\Big) \Big| n_j \in \mathbb{Z}_{>0}, \\ & n_1 + \cdots + n_r = n = 2m+1, \ k_j \in \mathbb{Z}, \ \frac{k_1}{n_1} > \cdots > \frac{k_{r-1}}{n_{r-1}} > 0\Big\} \end{aligned}$$

Recall that the twisted moduli spaces for $U(n)$ are defined by $\tilde{\mathcal{M}}^{\ell,i}_{n,k} = \tilde{V}^{\ell,i}_{n,k}/U(n)$, where $i = 1, 2$. Also we define the twisted moduli spaces for $SO(n)$ by

$$\mathcal{M}^{\ell,i,\pm 1}_{O(n),\pm 1} = V^{\ell,i,\pm 1}_{O(n),\pm 1}/SO(n), \quad \text{where } i = 1, 2.$$

PROPOSITION 6.8. *Suppose that* $\ell \geq 2i$, *where* $i = 1, 2$. *Let*

$$\mu = \Big(\underbrace{\frac{2k_1}{n_1}, \ldots, \frac{2k_1}{n_1}}_{n_1}, \ldots, \underbrace{\frac{2k_{r-1}}{n_{r-1}}, \ldots, \frac{2k_{r-1}}{n_{r-1}}}_{n_{r-1}}, \underbrace{0, \ldots, 0}_{n_r}\Big) \in \hat{I}_{SO(4m+2)}. \tag{6.7}$$

Then $X^{\ell,i}_{\rm YM}(SO(4m+2))_\mu$ *has two connected components (from both bundles over* Σ^ℓ_i*)*

$$X^{\ell,i}_{\rm YM}(SO(4m+2))^{+1}_\mu, \quad \text{and} \quad X^{\ell,i}_{\rm YM}(SO(4m+2))^{-1}_\mu.$$

We have a homeomorphism

$$\begin{aligned} &X^{\ell,i}_{\rm YM}(SO(4m+2))^{\pm1}_\mu / SO(4m+2) \\ \cong &\prod_{j=1}^{r-1} \tilde{\mathcal{M}}^{\ell,i}_{n_j,-k_j} \times \mathcal{M}^{\ell,i,\pm(-1)^{k_1+\cdots+k_{r-1}+im+i\frac{(n_r)(n_r-1)}{2}}}_{O(2n_r),(-1)^{n_r-1}} \end{aligned}$$

and a homotopy equivalence

$$\begin{aligned} &X^{\ell,i}_{\rm YM}(SO(4m+2))^{\pm1}_\mu{}^{hSO(4m+2)} \\ \sim &\prod_{j=1}^{r-1} \Big(\tilde{V}^{\ell,i}_{n_j,-k_j}\Big)^{hU(n_j)} \times \Big(V^{\ell,i,\pm(-1)^{k_1+\cdots+k_{r-1}+im+i\frac{(n_r)(n_r-1)}{2}}}_{O(2n_r),(-1)^{n_r-1}}\Big)^{hSO(2n_r)}. \end{aligned}$$

PROPOSITION 6.9. *Suppose that* $\ell \geq 2i$, *where* $i = 1, 2$. *The connected components of* $X^{\ell,i}_{\rm YM}(SO(4m+2))^{\pm1}$ *are*

$$\{X^{\ell,i}_{\rm YM}(SO(4m+2))^{\pm1}_\mu \mid \mu \in \hat{I}_{SO(4m+2)}\}.$$

Notice that, the set $\{\mu = \sqrt{-1}\mathrm{diag}(\mu_1 J, \ldots, \mu_{2m} J, 0J) \mid (\mu_1, \ldots, \mu_{2m}, 0) \in \hat{I}_{SO(4m+2)}\}$ is a *proper* subset of $\{\mu \in (\Xi^I_+)^\tau \mid I \subseteq \Delta, \tau(I) = I\}$ as mentioned in Section 4.5.

The following is an immediate consequence of Proposition 6.8.

THEOREM 6.10. *Suppose that* $\ell \geq 2i$, *where* $i = 1, 2$, *and let* μ *be as in* (6.7). *Then*

$$\begin{aligned} &P^{SO(4m+2)}_t \Big(X^{\ell,i}_{\rm YM}(SO(4m+2))^{\pm1}_\mu\Big) \\ = &\prod_{j=1}^{r-1} P^{U(n_j)}_t(\tilde{V}^{\ell,i}_{n_j,-k_j}) \cdot P^{SO(2n_r)}_t \Big(V^{\ell,i,\pm(-1)^{k_1+\cdots+k_{r-1}+im+i\frac{n_r(n_r-1)}{2}}}_{O(2n_r),(-1)^{n_r-1}}\Big). \end{aligned}$$

6.4. $SO(4m)$-connections on nonorientable surfaces

In this section, we consider $SO(2n)$ where $n = 2m$ is even, so that $\overline{C}_0^\tau = \overline{C}_0$. There are four cases.

Case 1. $t_{n-1} > |t_n|$, $n_r = 1$.

$$\mu = \sqrt{-1}\mathrm{diag}(\lambda_1 J_{n_1}, \ldots, \lambda_{r-1} J_{n_{r-1}}, \lambda_r J),$$

where $\lambda_1 > \cdots > \lambda_{r-1} > |\lambda_r| \geq 0$. Thus

$$SO(4m)_{X_\mu} \cong \Phi(U(n_1)) \times \cdots \times \Phi(U(n_{r-1})) \times \Phi(U(n_r)).$$

where $X_\mu = -2\pi\sqrt{-1}\mu$.

Let $\epsilon = H_{2m}$. Suppose that $(a_1, b_1, \ldots, a_\ell, b_\ell, \epsilon c', X_\mu/2) \in X^{\ell,1}_{\mathrm{YM}}(SO(4m))$. Then

$$\exp(X_\mu/2)\epsilon c'\epsilon c' = \prod_{i=1}^{\ell}[a_i, b_i]$$

where $a_i,\ b_i,\ c' \in \Phi(U(n_1)) \times \cdots \times \Phi(U(n_r))$.

Let $L : \mathbb{R}^{2n} \to \mathbb{C}^n$ be defined as in Section 5.1, and let

$$\begin{aligned} X'_\mu &= L \circ X_\mu \circ L^{-1} \\ &= 2\pi\sqrt{-1}\mathrm{diag}(\lambda_1 I_{n_1}, \cdots, \lambda_r I_{n_r}) \in \mathfrak{u}(n_1) \times \cdots \times \mathfrak{u}(n_r). \end{aligned}$$

Then the condition on X'_μ is

$$\exp(X'_\mu/2)\bar{c'}c' = \prod_{i=1}^{\ell}[a_i, b_i] \in SU(n_1) \times \cdots \times SU(n_{r-1}) \times \{I_2\}$$

where $a_i,\ b_i,\ c' \in U(n_1) \times \cdots \times U(n_r)$, and $\bar{c'}$ is the complex conjugate of c'. In order that this is nonempty, we need $1 = \det(e^{\pi\sqrt{-1}\lambda_j} I_{n_j})$, i.e.,

$$\lambda_j = \frac{2k_j}{n_j}, \quad k_j \in \mathbb{Z}, \quad j = 1, \ldots, r.$$

Similarly, suppose that $(a_1, b_1, \ldots, a_\ell, b_\ell, d, \epsilon c', X_\mu/2) \in X^{\ell,2}_{\mathrm{YM}}(SO(4m))$. Then

$$\exp(X_\mu/2)(\epsilon c')d(\epsilon c')^{-1}d = \prod_{i=1}^{\ell}[a_i, b_i]$$

where $a_i,\ b_i,\ d,\ c' \in \Phi(U(n_1)) \times \cdots \times \Phi(U(n_r))$. The condition on X'_μ is

$$\exp(X'_\mu/2)\bar{c'}\bar{d}\bar{c'}^{-1}d = \prod_{i=1}^{\ell}[a_i, b_i] \in SU(n_1) \times \cdots \times SU(n_{r-1}) \times \{I_2\},$$

where $a_i,\ b_i,\ d,\ c' \in U(n_1) \times \cdots \times U(n_r)$, and $\bar{c'}$ is the complex conjugate of c'. Again, we need

$$\lambda_j = \frac{2k_j}{n_j}, \quad k_j \in \mathbb{Z}, \quad j = 1, \ldots, r.$$

We conclude that for nonorientable surfaces

$$\mu = \sqrt{-1}\mathrm{diag}\Big(\frac{2k_1}{n_1}J_{n_1}, \ldots, \frac{2k_{r-1}}{n_{r-1}}J_{n_{r-1}}, 2k_r J\Big),\ k_j \in \mathbb{Z},\ \frac{k_1}{n_1} > \cdots > \frac{k_{r-1}}{n_{r-1}} > |k_r| \geq 0.$$

For each μ, define ϵ-reduced representation varieties

$$(6.8)\qquad \begin{aligned} V^{\ell,1}_{\mathrm{YM}}(SO(4m))_\mu = &\{(a_1, b_1, \ldots, a_\ell, b_\ell, c') \in SO(4m)^{2\ell+1}_{X_\mu} \mid \\ &\prod_{i=1}^{\ell}[a_i, b_i] = \exp(\frac{X_\mu}{2})\epsilon c'\epsilon c'\}, \end{aligned}$$

$$(6.9)\qquad \begin{aligned} V^{\ell,2}_{\mathrm{YM}}(SO(4m))_\mu = &\{(a_1, b_1, \ldots, a_\ell, b_\ell, d, c') \in SO(4m)^{2\ell+2}_{X_\mu} \mid \\ &\prod_{i=1}^{\ell}[a_i, b_i] = \exp(\frac{X_\mu}{2})\epsilon c' d(\epsilon c')^{-1}d\}. \end{aligned}$$

For $i = 1, \ldots, \ell$, write

$$a_i = \mathrm{diag}(A_1^i, \ldots, A_r^i), \quad b_i = \mathrm{diag}(B_1^i, \ldots, B_r^i),$$
$$c' = \mathrm{diag}(C_1, \ldots, C_r), \quad d = \mathrm{diag}(D_1, \ldots, D_r),$$

where $A_j^i,\ B_j^i,\ C_j,\ D_j \in \Phi(U(n_j))$.

Define V_j as in (5.7) when $i = 1$, and as in (5.9) when $i = 2$. Then $V_j \cong \tilde{V}^{\ell,i}_{n_j,-k_j}$ is connected, and

$$V^{\ell,1}_{\mathrm{YM}}(SO(4m))_\mu = \prod_{j=1}^r V_j.$$

Thus $V^{\ell,i}_{\mathrm{YM}}(SO(4m))_\mu$ is connected, and it corresponds to connections on a fixed topological $SO(4m)$-bundle P. By the argument in Section 5.3,

$$w_2(P) = k_1 + \cdots + k_r + i\frac{2m(2m+1)}{2} = k_1 + \cdots + k_r + im \pmod 2.$$

Let $U(n_j)$ acts on $V_j \cong \tilde{V}^{\ell,i}_{n_j,-k_j}$ by (4.9) and (4.10) in Section 4.6 when $i = 1$ and when $i = 2$, respectively. Then we have a homeomorphism

$$V^{\ell,i}_{\mathrm{YM}}(SO(4m))_\mu / SO(4m)_{X_\mu} \cong \prod_{j=1}^r (V_j/U(n_j)) \tag{6.10}$$

and a homotopy equivalence

$$V^{\ell,i}_{\mathrm{YM}}(SO(4m))_\mu^{hSO(4m)_{X_\mu}} \simeq \prod_{j=1}^r V_j^{hU(n_j)}. \tag{6.11}$$

Case 2. $t_{n-1} = -t_n > 0,\ n_r > 1.$

$$\mu = \sqrt{-1}\mathrm{diag}(\lambda_1 J_{n_1}, \ldots, \lambda_{r-1} J_{n_{r-1}}, \lambda_r J_{n_r-1}, -\lambda_r J),$$

where $\lambda_1 > \cdots > \lambda_r > 0$, Thus $SO(4m)_{X_\mu} \cong \Phi(U(n_1)) \times \cdots \times \Phi(U(n_{r-1})) \times \Phi'(U(n_r))$, where $\Phi : U(k) \hookrightarrow SO(2k)$ is the standard embedding, and $\Phi' : U(k) \hookrightarrow SO(2k)$ is defined as in Section 6.1.

Let $\epsilon = H_{2m}$. Suppose that $(a_1, b_1, \ldots, a_\ell, b_\ell, \epsilon c', X_\mu/2) \in X^{\ell,1}_{\mathrm{YM}}(SO(4m))$. Then

$$\exp(X_\mu/2)\epsilon c' \epsilon c' = \prod_{i=1}^\ell [a_i, b_i]$$

where $a_i,\ b_i,\ c' \in \Phi(U(n_1)) \times \cdots \times \Phi(U(n_{r-1}) \times \Phi'(U(n_r))$.

Let $L \oplus L' : \mathbb{R}^{2(n-n_r)} \oplus \mathbb{R}^{2n_r} \to \mathbb{C}^{n-n_r} \oplus \mathbb{C}^{n_r}$, and let

$$X'_\mu = (L \oplus L') \circ X_\mu \circ (L \otimes L')^{-1} = 2\pi\sqrt{-1}\mathrm{diag}(\lambda_1 I_{n_1}, \ldots, \lambda_r I_{n_r}) \in \mathfrak{u}(n_1) \times \cdots \times \mathfrak{u}(n_r).$$

Then the condition on X'_μ is

$$\exp(X'_\mu/2)\bar{c'}c' = \prod_{i=1}^\ell [a_i, b_i] \in SU(n_1) \times \cdots \times SU(n_r)$$

where $a_i,\ b_i,\ c' \in U(n_1) \times \cdots \times U(n_r)$, and $\bar{c'}$ is the complex conjugate of c'. In order that this is nonempty, we need $1 = \det(e^{\pi\sqrt{-1}\lambda_j} I_{n_j})$, i.e.,

$$\lambda_j = \frac{2k_j}{n_j}, \quad k_j \in \mathbb{Z}, \quad j = 1, \ldots, r.$$

Similarly, suppose that $(a_1, b_1, \ldots, a_\ell, b_\ell, d, \epsilon c', X_\mu/2) \in X_{\mathrm{YM}}^{\ell,2}(SO(4m))$. Then

$$\exp(X_\mu/2)(\epsilon c')d(\epsilon c')^{-1}d = \prod_{i=1}^{\ell}[a_i, b_i],$$

where $a_i,\ b_i,\ d,\ c' \in \Phi(U(n_1)) \times \cdots \times \Phi(U(n_{r-1})) \times \Phi'(U(n_r))$. The condition on X'_μ is

$$\exp(X'_\mu/2)\bar{c}'\bar{d}\bar{c}'^{-1}d = \prod_{i=1}^{\ell}[a_i, b_i] \in SU(n_1) \times \cdots \times SU(n_r),$$

where $a_i,\ b_i,\ d,\ c' \in U(n_1) \times \cdots \times U(n_r)$ and $\bar{d}$ is the complex conjugate of d.

Again, we need

$$\lambda_j = \frac{2k_j}{n_j}, \quad k_j \in \mathbb{Z}, \quad j = 1, \ldots, r.$$

We conclude that for nonorientable surfaces

$$\mu = \sqrt{-1}\mathrm{diag}(\frac{2k_1}{n_1}J_{n_1}, \ldots, \frac{2k_{r-1}}{n_{r-1}}J_{n_{r-1}}, \frac{2k_r}{n_r}J_{n_r-1}, -\frac{2k_r}{n_r}J),\ k_j \in \mathbb{Z},\ \frac{k_1}{n_1} > \cdots > \frac{k_r}{n_r} > 0.$$

For each μ, define ϵ-reduced representation varieties as in (6.8) and (6.9). For $i = 1, \ldots, \ell$, write

$$a_i = \mathrm{diag}(A_1^i, \ldots, A_r^i), \quad b_i = \mathrm{diag}(B_1^i, \ldots, B_r^i),$$
$$c' = \mathrm{diag}(C_1, \ldots, C_r), \quad d = \mathrm{diag}(D_1, \ldots, D_r),$$

where $A_j^i,\ B_j^i,\ C_j,\ D_j \in \Phi(U(n_j))$ for $j = 1, \cdots, r-1$, and $A_r^i,\ B_r^i,\ C_r,\ D_r \in \Phi'(U(n_r))$.

$i = 1$. For $j = 1, \ldots, r-1$, define V_j as in (5.7). Define

$$\begin{aligned} V_r &\overset{\Phi'}{\cong} \Big\{(A_r^1, B_r^1, \ldots, A_r^\ell, B_r^\ell, C_r) \in U(n_r)^{2\ell+1} \mid \prod_{i=1}^{\ell}[A_r^i, B_r^i] = e^{\frac{2\pi\sqrt{-1}k_r}{n_r}} I_{n_r}\bar{C}_r C_r\Big\} \\ &\cong \tilde{V}_{n_r,-k_r}^{\ell,1}. \end{aligned}$$

Then $V_{\mathrm{YM}}^{\ell,1}(SO(4m))_\mu = \prod_{j=1}^{r} V_j$.

$i = 2$. For $j = 1, \ldots, r-1$, define V_j as in (5.9). Define

$$\begin{aligned} V_r = \Big\{&(A_r^1, B_r^1, \ldots, A_r^\ell, B_r^\ell, D_r, C_r) \in U(n_r)^{2\ell+2} \mid \\ &\prod_{i=1}^{\ell}[A_r^i, B_r^i] = e^{\frac{2\pi\sqrt{-1}k_r}{n_r}} I_{n_r}\bar{C}_r\bar{D}_r\bar{C}_r^{-1}D_r\Big\} \\ \cong\ & \tilde{V}_{n_r,-k_r}^{\ell,2}. \end{aligned}$$

Then $V_{\mathrm{YM}}^{\ell,2}(SO(4m))_\mu = \prod_{j=1}^{r} V_j$.

Thus $V_{\mathrm{YM}}^{\ell,i}(SO(4m))_\mu$ is also connected, so it corresponds to a fixed topological $SO(4m)$-bundle P. As in Case 1,

$$w_2(P) = k_1 + \cdots + k_r + im \pmod 2.$$

We also have a homeomorphism (6.10) and a homotopy equivalence (6.11).

Case 3. $t_{n-1} = t_n > 0$, $n_r > 1$.

$$\mu = \sqrt{-1}\mathrm{diag}(\lambda_1 J_{n_1}, \ldots, \lambda_r J_{n_r}),$$

where $\lambda_1 > \cdots > \lambda_r > 0$. Let $X_\mu = -2\pi\sqrt{-1}\mu$ as before. Then

$$SO(2n)_\mu = SO(2n)_{X_\mu} \cong \Phi(U(n_1)) \times \cdots \times \Phi(U(n_r)).$$

Let $\epsilon = H_{2m}$ as in Example 4.11. Suppose that $(a_1, b_1, \ldots, a_\ell, b_\ell, \epsilon c', X_\mu/2) \in X^{\ell,1}_{\mathrm{YM}}(SO(4m))$. Then

$$\exp(X_\mu/2)\epsilon c'\epsilon c' = \prod_{i=1}^{\ell}[a_i, b_i]$$

where a_i, b_i, $c' \in \Phi(U(n_1)) \times \cdots \times \Phi(U(n_r))$.

Let $L : \mathbb{R}^{2n} \to \mathbb{C}^n$ be defined as in Section 5.1, and let

$$\begin{aligned} X'_\mu &= L \circ X_\mu \circ L^{-1} \\ &= 2\pi\sqrt{-1}\mathrm{diag}(\lambda_1 I_{n_1}, \ldots, \lambda_r I_{n_r}) \in \mathfrak{u}(n_1) \times \cdots \times \mathfrak{u}(n_r). \end{aligned}$$

Then the condition on X'_μ is

$$\exp(X'_\mu/2)\bar{c'}c' = \prod_{i=1}^{\ell}[a_i, b_i] \in SU(n_1) \times \cdots \times SU(n_r),$$

where a_i, b_i, $c' \in U(n_1) \times \cdots \times U(n_r)$, and $\bar{c'}$ is the complex conjugate of c'. In order that this is nonempty, we need $1 = \det(e^{\pi\sqrt{-1}\lambda_j} I_{n_j})$, i.e.,

$$\lambda_j = \frac{2k_j}{n_j}, \quad k_j \in \mathbb{Z}, \quad j = 1, \ldots, r.$$

Similarly, suppose that $(a_1, b_1, \ldots, a_\ell, b_\ell, d, \epsilon c', X_\mu/2) \in X^{\ell,2}_{\mathrm{YM}}(SO(4m))$. Then

$$\exp(X_\mu/2)(\epsilon c')d(\epsilon c')^{-1}d = \prod_{i=1}^{\ell}[a_i, b_i]$$

where a_i, b_i, d, $c' \in \Phi(U(n_1)) \times \cdots \times \Phi(U(n_r))$. The condition on X'_μ is

$$\exp(X'_\mu/2)\bar{c'}\bar{d}\bar{c'}^{-1}d \in SU(n_1) \times \cdots \times SU(n_r),$$

where d, $c' \in U(n_1) \times \cdots \times U(n_r)$, and $\bar{d}$ is the complex conjugate of d. Again, we need

$$\lambda_j = \frac{2k_j}{n_j}, \quad k_j \in \mathbb{Z}, \quad j = 1, \ldots, r.$$

We conclude that for nonorientable surfaces,

$$\mu = \sqrt{-1}\mathrm{diag}\Big(\frac{2k_1}{n_1} J_{n_1}, \ldots, \frac{2k_r}{n_r} J_{n_r}\Big), \quad k_j \in \mathbb{Z}, \quad \frac{k_1}{n_1} > \cdots \frac{k_r}{n_r} > 0.$$

For each μ, we define the ϵ-reduced representation varieties as in (6.8) and (6.9) when $i = 1$ and when $i = 2$, respectively; we define V_j as in (5.7) and (5.9) when $i = 1$ and when $i = 2$, respectively. Then

$$V^{\ell,i}_{\mathrm{YM}}(SO(4m))_\mu = \prod_{j=1}^{r} V_j.$$

Again, $V^{\ell,i}_{\mathrm{YM}}(SO(4m))_\mu$ is connected, so it corresponds to a fixed topological $SO(4m)$-bundle P, and

$$w_2(P) = k_1 + \cdots + k_r + im \pmod 2.$$

We also have a homeomorphism (6.10) and a homotopy equivalence (6.11).

Case 4. $t_{n-1} = t_n = 0$, $n_r > 1$.

$$\mu = \sqrt{-1}\mathrm{diag}(\lambda_1 J_{n_1}, \ldots, \lambda_{r-1} J_{n_{r-1}}, 0 J_{n_r}),$$

where $\lambda_1 > \cdots > \lambda_{r-1} > 0$. Let $X_\mu = -2\pi\sqrt{-1}\mu$ as before. Then

$$SO(2n)_\mu = SO(2n)_{X_\mu} \cong \Phi(U(n_1)) \times \cdots \times \Phi(U(n_{r-1})) \times SO(2n_r).$$

Let $\epsilon_\mu = \mathrm{diag}(H_{2m-n_r}, (-1)^{n_r} I_1, I_{2n_r-1})$. Consider $(a_1, b_1, \ldots, a_\ell, b_\ell, \epsilon_\mu c', X_\mu/2) \in X^{\ell,1}_{\mathrm{YM}}(SO(4m))$. Then

$$\exp(X_\mu/2)\epsilon_\mu c' \epsilon_\mu c' = \prod_{i=1}^{\ell}[a_i, b_i]$$

where $a_i,\ b_i,\ c' \in \Phi(U(n_1)) \times \cdots \times \Phi(U(n_{r-1})) \times SO(2n_r)$.

Let $L : \mathbb{R}^{2(n-n_r)} \to \mathbb{C}^{n-n_r}$ be defined as in Section 5.1, and let

$$\begin{aligned} X'_\mu &= L \circ \left(2\pi\mathrm{diag}(\lambda_1 J_{n_1}, \ldots, \lambda_{r-1} J_{n_{r-1}})\right) \circ L^{-1} \\ &= 2\pi\sqrt{-1}\mathrm{diag}(\lambda_1 I_{n_1}, \ldots, \lambda_{r-1} I_{n_{r-1}}) \in \mathfrak{u}(n_1) \times \cdots \times \mathfrak{u}(n_{r-1}). \end{aligned}$$

Then the condition on X'_μ is

$$\exp(X'_\mu/2)\bar{c'}c' = \prod_{i=1}^{\ell}[a_i, b_i] \in SU(n_1) \times \cdots \times SU(n_{r-1}),$$

where $a_i,\ b_i,\ c' \in U(n_1) \times \cdots \times U(n_{r-1})$, and $\bar{c'}$ is the complex conjugate of c'. In order that this is nonempty, we need $1 = \det(e^{\pi\sqrt{-1}\lambda_j} I_{n_j})$, i.e.,

$$\lambda_j = \frac{2k_j}{n_j}, \quad k_j \in \mathbb{Z}, \quad j = 1, \ldots, r-1.$$

Similarly, suppose that $(a_1, b_1, \ldots, a_\ell, b_\ell, d, \epsilon_\mu c', X_\mu/2) \in X^{\ell,2}_{\mathrm{YM}}(SO(4m))$. Then

$$\exp(X_\mu/2)(\epsilon_\mu c')d(\epsilon_\mu c')^{-1}d = \prod_{i=1}^{\ell}[a_i, b_i],$$

where $a_i,\ b_i,\ d,\ c' \in \Phi(U(n_1)) \times \cdots \times \Phi(U(n_{r-1})) \times SO(2n_r)$. The condition on X'_μ is

$$\exp(X'_\mu/2)\bar{c'}\bar{d}\bar{c'}^{-1}d \in SU(n_1) \times \cdots \times SU(n_{r-1}),$$

where $d,\ c' \in U(n_1) \times \cdots \times U(n_{r-1})$, and $\bar{d}$ is the complex conjugate of d. Again, we need

$$\lambda_j = \frac{2k_j}{n_j}, \quad k_j \in \mathbb{Z}, \quad j = 1, \ldots, r-1.$$

We conclude that for nonorientable surfaces,

$$\mu = \sqrt{-1}\mathrm{diag}\Big(\frac{2k_1}{n_1} J_{n_1}, \ldots, \frac{2k_{r-1}}{n_{r-1}} J_{n_{r-1}}, 0 J_{n_r}\Big), \quad k_j \in \mathbb{Z}, \quad \frac{k_1}{n_1} > \cdots > \frac{k_{r-1}}{n_{r-1}} > 0.$$

For each μ, define ϵ_μ-reduced representation varieties

$$\begin{aligned} V^{\ell,1}_{\mathrm{YM}}(SO(4m))_\mu &= \{(a_1,b_1,\ldots,a_\ell,b_\ell,c') \in SO(4m)^{2\ell+1}_{X_\mu} \mid \\ &\prod_{i=1}^{\ell}[a_i,b_i] = \exp(X_\mu/2)\epsilon_\mu c'\epsilon_\mu c'\}, \\ V^{\ell,2}_{\mathrm{YM}}(SO(4m))_\mu &= \{(a_1,b_1,\ldots,a_\ell,b_\ell,d,c') \in SO(4m)^{2\ell+2}_{X_\mu} \mid \\ &\prod_{i=1}^{\ell}[a_i,b_i] = \exp(X_\mu/2)\epsilon_\mu c' d(\epsilon_\mu c')^{-1}d\}. \end{aligned}$$

$i=1$. For $j=1,\ldots,r-1$, define V_j as in (5.7). Define

$$V_r = \Big\{(A^1_r,B^1_r,\ldots,A^\ell_r,B^\ell_r,C_r) \in SO(2n_r)^{2\ell+1} \mid \prod_{i=1}^{\ell}[A^i_r,B^i_r] = (\epsilon C_r)^2\Big\}, \tag{6.12}$$

where $\epsilon = \mathrm{diag}((-1)^{n_r}I_1, I_{2n_r-1})$, $\det(\epsilon) = (-1)^{n_r}$. Let $C_0' = \epsilon C_0$. We see that

$$\begin{aligned} V_r &\cong \Big\{(A^1_r,B^1_r,\ldots,A^\ell_r,B^\ell_r,C'_r) \in SO(2n_r)^{2\ell}\times O(2n_r) \mid \\ &\prod_{i=1}^{\ell}[A^i_r,B^i_r] = (C'_r)^2, \det(C'_r) = (-1)^{n_r}\Big\} \\ &\cong V^{\ell,1}_{O(2n_r),(-1)^{n_r}} \end{aligned}$$

where $V^{\ell,1}_{O(n),\pm1}$ is the twisted representation variety defined in (4.11) of Section 4.7. $V^{\ell,1}_{O(n),\pm1}$ is nonempty if $\ell \geq 2$. We have shown that $V^{\ell,1}_{O(n),\pm1}$ is disconnected with two components $V^{\ell,1,+1}_{O(n),\pm1}$ and $V^{\ell,1,-1}_{O(n),\pm1}$ if $\ell \geq 2$ and $n > 2$ (Proposition 4.14). Then

$$V^{\ell,1}_{\mathrm{YM}}(SO(4m))_\mu = \prod_{j=1}^{r} V_j.$$

$i=2$. For $j=1,\ldots,r-1$, define V_j as in (5.9). Define

$$V_r = \Big\{(A^1_r,B^1_r,\ldots,A^\ell_r,B^\ell_r,D_r,C_r) \in SO(2n_r)^{2\ell+2} \mid \prod_{i=1}^{\ell}[A^i_r,B^i_r] = \epsilon C_r D_r(\epsilon C_r)^{-1}D_r\Big\}, \tag{6.13}$$

where $\epsilon = \mathrm{diag}((-1)^{n_r}I_1, I_{2n_r-1})$, $\det(\epsilon) = (-1)^{n_r}$. Let $C'_r = \epsilon C_r$. We see that

$$\begin{aligned} V_r &\cong \Big\{(A^1_r,B^1_r,\ldots,A^\ell_r,B^\ell_r,D_r,C'_r) \in SO(2n_r)^{2\ell+1}\times O(2n_r) \mid \\ &\prod_{i=1}^{\ell}[A^i_r,B^i_r] = C'_r D_r C'^{-1}_r D_r, \det(C'_r) = (-1)^{n_r}\Big\} \\ &\cong V^{\ell,2}_{O(2n_r),(-1)^{n_r}} \end{aligned}$$

where $V^{\ell,2}_{O(n),\pm1}$ is the twisted representation variety defined in (4.12) of Section 4.7. $V^{\ell,2}_{O(n),\pm1}$ is nonempty if $\ell \geq 4$. We have shown that $V^{\ell,2}_{O(n),\pm1}$ is disconnected with two components $V^{\ell,2,+1}_{O(n),\pm 1}$ and $V^{\ell,2,-1}_{O(n),\pm 1}$ if $\ell \geq 4$ and $n > 2$ (Proposition 4.14). Then

$$V^{\ell,2}_{\mathrm{YM}}(SO(4m))_\mu = \prod_{j=1}^{r} V_j.$$

Thus $V^{\ell,i}_{\mathrm{YM}}(SO(4m))_\mu$ is disconnected with two connected components if $\ell \geq 2i$ and $n_r > 1$ (because V_r is). By the argument in Section 6.3,

$$V^{\ell,i}_{\mathrm{YM}}(SO(4m))^{\pm 1}_\mu = \prod_{j=1}^{r-1} V_j \times V^{\ell,i,\pm(-1)^{k_1+\cdots+k_{r-1}+i\frac{(n-n_r)(n-n_r-1)}{2}}}_{O(2n_r),(-1)^{n_r}}.$$

Note that $i\dfrac{(n-n_r)(n-n_r-1)}{2} \equiv i(m + \dfrac{n_r(n_r+1)}{2})$.

Let $U(n_j)$ act on $V_j = \tilde{V}^{\ell,i}_{n_j,-k_j}$ by (4.9) and (4.10) of Section 4.6 when $i=1$ and when $i=2$, respectively; let $SO(2n_r)$ act on $V_r = V^{\ell,i}_{O(2n_r),(-1)^{n_r}}$ by (4.13) and (4.14) in Section 4.7 when $i=1$ and $i=2$, respectively. Then we have a homeomorphism

$$V^{\ell,i}_{\mathrm{YM}}(SO(4m))_\mu / SO(4m)_{X_\mu} \cong \prod_{j=1}^{r-1} (V_j/U(n_j)) \times V_r/SO(2n_r),$$

and a homotopy equivalence

$$V^{\ell,i}_{\mathrm{YM}}(SO(4m))_\mu{}^{hSO(4m)_{X_\mu}} \sim \prod_{j=1}^{r-1} V_j{}^{hU(n_j)} \times V_r{}^{hSO(2n_r)}.$$

To simplify the notation, we write

$$\mu = (\mu_1, \ldots, \mu_{2m}) = \Big(\underbrace{\frac{2k_1}{n_1}, \ldots, \frac{2k_1}{n_1}}_{n_1}, \ldots, \underbrace{\frac{2k_r}{n_r}, \ldots, \frac{2k_r}{n_r}}_{n_r}\Big)$$

instead of

$$\sqrt{-1}\mathrm{diag}\Big(\frac{2k_1}{n_1}J_{n_1}, \ldots, \frac{2k_r}{n_r}J_{n_r}\Big).$$

Let

$$\begin{aligned}\hat{I}^{\pm 1}_{SO(4m)} = \Big\{\mu = \Big(&\underbrace{\frac{2k_1}{n_1}, \ldots, \frac{2k_1}{n_1}}_{n_1}, \ldots, \underbrace{\frac{2k_{r-1}}{n_{r-1}}, \ldots, \frac{2k_{r-1}}{n_{r-1}}}_{n_{r-1}}, 2k_r\Big) \Big| n_j \in \mathbb{Z}_{>0},\ k_j \in \mathbb{Z}, \\ &n_1 + \cdots + n_{r-1} + 1 = n,\ \frac{k_1}{n_1} > \cdots > \frac{k_{r-1}}{n_{r-1}} > |k_r|,\ (-1)^{k_1+\cdots+k_r+im} = \pm 1\Big\} \\ \bigcup \Big\{\mu = \Big(&\underbrace{\frac{2k_1}{n_1}, \ldots, \frac{2k_1}{n_1}}_{n_1}, \ldots, \underbrace{\frac{2k_{r-1}}{n_{r-1}}, \ldots, \frac{2k_{r-1}}{n_{r-1}}}_{n_{r-1}}, \underbrace{\frac{2k_r}{n_r}, \ldots, \frac{2k_r}{n_r}}_{n_r-1}, \pm\frac{2k_r}{n_r}\Big) \Big| \ n_j \in \mathbb{Z}_{>0}, \\ &n_r > 1,\ n_1 + \cdots + n_r = n,\ k_j \in \mathbb{Z},\ \frac{k_1}{n_1} > \cdots > \frac{k_r}{n_r} > 0,\ (-1)^{k_1+\cdots+k_r+im} = \pm 1\Big\},\end{aligned}$$

$$\hat{I}^0_{SO(4m)} = \Big\{\mu = \Big(\underbrace{\frac{2k_1}{n_1},\ldots,\frac{2k_1}{n_1}}_{n_1},\ldots,\underbrace{\frac{2k_{r-1}}{n_{r-1}},\ldots,\frac{2k_{r-1}}{n_{r-1}}}_{n_{r-1}},\underbrace{0,\ldots,0}_{n_r}\Big)\,\Big|\, n_j\in\mathbb{Z}_{>0},$$

$$n_r>1,\ n_1+\cdots+n_r=n,\ k_j\in\mathbb{Z},\ \frac{k_1}{n_1}>\cdots>\frac{k_{r-1}}{n_{r-1}}>0\Big\}.$$

PROPOSITION 6.11. *Suppose that $\ell\geq 2i$, where $i=1,2$.*

(i) *If* $\mu = \Big(\underbrace{\frac{2k_1}{n_1},\ldots,\frac{2k_1}{n_1}}_{n_1},\ldots,\underbrace{\frac{2k_{r-1}}{n_{r-1}},\ldots,\frac{2k_{r-1}}{n_{r-1}}}_{n_{r-1}},2k_r\Big)\in\hat{I}^{\pm1}_{SO(4m)}$, *or*

$$\mu = \Big(\underbrace{\frac{2k_1}{n_1},\ldots,\frac{2k_1}{n_1}}_{n_1},\ldots,\underbrace{\frac{2k_{r-1}}{n_{r-1}},\ldots,\frac{2k_{r-1}}{n_{r-1}}}_{n_{r-1}},\underbrace{\frac{2k_r}{n_r},\ldots,\frac{2k_r}{n_r}}_{n_r-1},\pm\frac{2k_r}{n_r}\Big)\in\hat{I}^{\pm1}_{SO(4m)},$$

then

$$X^{\ell,i}_{\mathrm{YM}}(SO(4m))_\mu = X^{\ell,i}_{\mathrm{YM}}(SO(4m))^{\pm1}_\mu$$

is nonempty and connected. We have a homeomorphism

$$X^{\ell,i}_{\mathrm{YM}}(SO(4m))_\mu/SO(4m)\cong\prod_{j=1}^r\tilde{\mathcal{M}}^{\ell,i}_{n_j,-k_j}$$

and a homotopy equivalence

$$X^{\ell,i}_{\mathrm{YM}}(SO(4m))_\mu{}^{hSO(4m)}\sim\prod_{j=1}^r\big(\tilde{V}^{\ell,i}_{n_j,-k_j}\big)^{hU(n_j)}.$$

(ii) *If* $\mu = \Big(\underbrace{\frac{2k_1}{n_1},\ldots,\frac{2k_1}{n_1}}_{n_1},\ldots,\underbrace{\frac{2k_{r-1}}{n_{r-1}},\ldots,\frac{2k_{r-1}}{n_{r-1}}}_{n_{r-1}},\underbrace{0,\ldots,0}_{n_r}\Big)\in\hat{I}^0_{SO(4m)}$,

then $X^{\ell,i}_{\mathrm{YM}}(SO(4m))_\mu$ has two connected components (from both bundles over Σ^ℓ_i)

$$X^{\ell,i}_{\mathrm{YM}}(SO(4m))^{+1}_\mu,\quad\text{and}\quad X^{\ell,i}_{\mathrm{YM}}(SO(4m))^{-1}_\mu.$$

We have homeomorphisms

$$X^{\ell,i}_{\mathrm{YM}}(SO(4m))^{\pm1}_\mu/SO(4m)\cong\prod_{j=1}^{r-1}\tilde{\mathcal{M}}^{\ell,i}_{n_j,-k_j}\times\mathcal{M}^{\ell,i,\pm(-1)^{k_1+\cdots+k_{r-1}+im+i\frac{n_r(n_r+1)}{2}}}_{O(2n_r),(-1)^{n_r}}$$

and homotopy equivalences

$$\Big(X^{\ell,i}_{\mathrm{YM}}(SO(4m))^{\pm1}_\mu\Big)^{hSO(4m)}$$
$$\sim\ \prod_{j=1}^{r-1}\big(\tilde{V}^{\ell,i}_{n_j,-k_j}\big)^{hU(n_j)}\times\Big(\tilde{V}^{\ell,i,\pm(-1)^{k_1+\cdots+k_{r-1}+im+i\frac{n_r(n_r+1)}{2}}}_{O(2n_r),(-1)^{n_r}}\Big)^{hSO(2n_r)}.$$

PROPOSITION 6.12. *Suppose that $\ell\geq 2i$, where $i=1,2$. The connected components of $X^{\ell,i}_{\mathrm{YM}}(SO(4m))^{\pm1}$ are*

$$\{X^{\ell,i}_{\mathrm{YM}}(SO(4m))_\mu\mid\mu\in\hat{I}^{\pm1}_{SO(4m)}\}\cup\{X^{\ell,i}_{\mathrm{YM}}(SO(4m))^{\pm1}_\mu\mid\mu\in\hat{I}^0_{SO(4m)}\}.$$

Notice that, the set $\{\mu = \sqrt{-1}\mathrm{diag}(\mu_1 J, \ldots, \mu_{2m} J) \mid (\mu_1, \ldots, \mu_{2m}) \in \hat{I}^{\pm 1}_{SO(4m)} \cup \hat{I}^{0}_{SO(4m)}\}$ is a *proper* subset of $\{\mu \in (\Xi^I_+)^\tau \mid I \subseteq \Delta, \tau(I) = I\}$ as mentioned in Section 4.5.

The following is an immediate consequence of Proposition 6.11.

PROPOSITION 6.13. *Suppose that* $\ell \geq 2i$, *where* $i = 1, 2$.

(i) *If* $\mu = \Big(\underbrace{\frac{2k_1}{n_1}, \ldots, \frac{2k_1}{n_1}}_{n_1}, \ldots, \underbrace{\frac{2k_{r-1}}{n_{r-1}}, \ldots, \frac{2k_{r-1}}{n_{r-1}}}_{n_{r-1}}, 2k_r\Big) \in \hat{I}^{\pm 1}_{SO(4m)}$, *or*

$$\mu = \Big(\underbrace{\frac{2k_1}{n_1}, \ldots, \frac{2k_1}{n_1}}_{n_1}, \ldots, \underbrace{\frac{2k_{r-1}}{n_{r-1}}, \ldots, \frac{2k_{r-1}}{n_{r-1}}}_{n_{r-1}}, \underbrace{\frac{2k_r}{n_r}, \ldots, \frac{2k_r}{n_r}}_{n_r - 1}, \pm\frac{2k_r}{n_r}\Big) \in \hat{I}^{\pm 1}_{SO(4m)},$$

then

$$P_t^{SO(4m)}\left(X^{\ell,i}_{\mathrm{YM}}(SO(4m))_\mu\right) = \prod_{j=1}^{r} P_t^{U(n_j)}(\tilde{V}^{\ell,i}_{n_j,-k_j}).$$

(ii) *If* $\mu = \Big(\underbrace{\frac{2k_1}{n_1}, \ldots, \frac{2k_1}{n_1}}_{n_1}, \ldots, \underbrace{\frac{2k_{r-1}}{n_{r-1}}, \ldots, \frac{2k_{r-1}}{n_{r-1}}}_{n_{r-1}}, \underbrace{0, \ldots, 0}_{n_r}\Big) \in \hat{I}^{0}_{SO(4m)}$, *then*

$$\begin{aligned} & P_t^{SO(4m)}\left(X^{\ell,i}_{\mathrm{YM}}(SO(4m))^{\pm 1}_\mu\right) \\ = & \prod_{j=1}^{r-1} P_t^{U(n_j)}(\tilde{V}^{\ell,i}_{n_j,-k_j}) \times P_t^{SO(2n_r)}\left(V^{\ell,i,\pm(-1)^{k_1+\cdots+k_{r-1}+im+i\frac{n_r(n_r+1)}{2}}}_{O(2n_r),(-1)^{n_r}}\right). \end{aligned}$$

CHAPTER 7

Yang-Mills $Sp(n)$-Connections

$$Sp(n) = \left\{ \begin{pmatrix} A & -\bar{B} \\ B & \bar{A} \end{pmatrix} \in U(2n) \;\middle|\; A, B \in GL(n, \mathbb{C}) \right\}$$

The maximal torus of $Sp(n)$ consists of diagonal matrices of the form

$$\mathrm{diag}(u_1, \ldots, u_n, u_1^{-1}, \ldots, u_n^{-1}),$$

where $u_1, \ldots, u_n \in U(1)$. The Lie algebra of the maximal torus consists of diagonal matrices of the form

$$-2\pi\sqrt{-1}\mathrm{diag}(t_1, \ldots, t_n, -t_1, \ldots, -t_n), \quad t_i \in \mathbb{R}.$$

The fundamental Weyl chamber is

$$\overline{C}_0 = \{\mathrm{diag}(t_1, \ldots, t_n, -t_1, \ldots, -t_n) \mid t_1 \geq t_2 \geq \cdots \geq t_n \geq 0\}.$$

In this chapter, we assume

$$n_1, \ldots, n_r \in \mathbb{Z}_{>0}, \quad n_1 + \cdots + n_r = n.$$

7.1. $Sp(n)$-connections on orientable surfaces

Any $\mu \in \overline{C}_0$ is of the form

$$\mu = \mathrm{diag}(\lambda_1 I_{n_1}, \ldots, \lambda_r I_{n_r}, -\lambda_1 I_{n_1}, \ldots, -\lambda_r I_{n_r}),$$

where $\lambda_1 > \cdots > \lambda_r \geq 0$. When $\lambda_r > 0$, $Sp(n)_{X_\mu}$ consists of matrices of the form

$$\mathrm{diag}(M_1, \ldots, M_r, \overline{M}_1, \ldots, \overline{M}_r),$$

where $M_j \in U(n_j)$. When $\lambda_r = 0$, $Sp(n)_{X_\mu}$ consists of matrices of the form

$$\begin{pmatrix} M_1 & & & & & & 0 \\ & \cdot & & & & & \\ & & M_{r-1} & & & & \\ & & & M_r & 0 & & -\overline{N}_r \\ & & & 0 & \overline{M}_1 & & \\ & & & & & \cdot & \\ & & & & & \overline{M}_{r-1} & \\ 0 & & & N_r & & & \overline{M}_r \end{pmatrix}$$

where $M_j \in U(n_j)$ for $j = 1, \ldots, r-1$, and

$$S = \begin{pmatrix} M_r & -\overline{N}_r \\ N_r & \overline{M}_r \end{pmatrix} \in Sp(n_r) \subset U(2n_r).$$

So

$$Sp(n)_{X_\mu} \cong \begin{cases} U(n_1) \times \cdots \times U(n_r), & \lambda_r > 0, \\ U(n_1) \times \cdots \times U(n_{r-1}) \times Sp(n_r), & \lambda_r = 0. \end{cases}$$

Suppose that $(a_1, b_1, \ldots, a_\ell, b_\ell, X_\mu) \in X_{\mathrm{YM}}^{\ell,0}(Sp(n))$. Then

$$\exp(X_\mu) = \prod_{i=1}^{\ell}[a_i, b_i],$$

where $a_1, b_1, \ldots, a_\ell, b_\ell \in Sp(n)_{X_\mu}$. Then we have

$$\exp(X_\mu) \in (Sp(n)_{X_\mu})_{ss} \cong \begin{cases} SU(n_1) \times \cdots \times SU(n_r), & \lambda_r > 0, \\ SU(n_r) \times \cdots \times SU(n_{r-1}) \times Sp(n_r), & \lambda_r = 0. \end{cases}$$

Thus

$$\begin{aligned} X_\mu &= -2\pi\sqrt{-1}\mathrm{diag}\Big(\frac{k_1}{n_1}I_{n_1}, \ldots, \frac{k_r}{n_r}I_{n_r}, -\frac{k_1}{n_1}I_{n_1}, \ldots, -\frac{k_r}{n_r}I_{n_r}\Big), \\ \mu &= \mathrm{diag}\Big(\frac{k_1}{n_1}I_{n_1}, \ldots, \frac{k_r}{n_r}I_{n_r}, -\frac{k_1}{n_1}I_{n_1}, \ldots, -\frac{k_r}{n_r}I_{n_r}\Big), \end{aligned}$$

where

$$k_j \in \mathbb{Z}, \quad \frac{k_1}{n_1} > \cdots > \frac{k_r}{n_r} \geq 0.$$

This agrees with Section 3.4.4.

Recall for each μ, the representation variety is

$$V_{\mathrm{YM}}^{\ell,0}(Sp(n))_\mu = \{(a_1, b_1, \ldots, a_\ell, b_\ell) \in (Sp(n)_{X_\mu})^{2\ell} \mid \prod_{i=1}^{\ell}[a_i, b_i] = \exp(X_\mu)\}.$$

Let $i = 1, \ldots, \ell$. When $k_r > 0$, write

$$a_i = \mathrm{diag}\left(A_1^i, \ldots, A_r^i, \bar{A}_1^i, \ldots, \bar{A}_r^i\right), \quad b_i = \mathrm{diag}\left(B_1^i, \ldots, B_r^i, \bar{B}_1^i, \ldots, \bar{B}_r^i\right),$$

where $A_j^i, B_j^i \in U(n_j)$. When $k_r = 0$, write

$$a_i = \begin{pmatrix} A^i & & & 0 \\ & A_r^i & & -\bar{E}_r^i \\ & & \bar{A}^i & \\ 0 & E_r^i & & \bar{A}_r^i \end{pmatrix}, \quad b_i = \begin{pmatrix} B^i & & & 0 \\ & B_r^i & & -\bar{F}_r^i \\ & & \bar{B}^i & \\ 0 & F_r^i & & \bar{B}_r^i \end{pmatrix},$$

where

$$\begin{aligned} &A^i = \mathrm{diag}\left(A_1^i, \ldots, A_{r-1}^i\right), \quad B^i = \mathrm{diag}\left(B_1^i, \ldots, B_{r-1}^i\right), \\ &A_j^i, B_j^i \in U(n_j), \quad j = 1, \ldots, r-1, \\ &P^i = \begin{pmatrix} A_r^i & -\bar{E}_r^i \\ E_r^i & \bar{A}_r^i \end{pmatrix}, Q^i = \begin{pmatrix} B_r^i & -\bar{F}_r^i \\ F_r^i & \bar{B}_r^i \end{pmatrix} \in Sp(n_r) \subset U(2n_r). \end{aligned}$$

For $j = 1, \ldots, r-1$, define

$$\begin{aligned} V_j &= \Big\{(A_j^1, B_j^1, \ldots, A_j^\ell, B_j^\ell) \in U(n_j)^{2\ell} \mid \prod_{i=1}^{\ell}[A_j^i, B_j^i] = e^{-2\pi\sqrt{-1}\frac{k_j}{n_j}} I_{n_j}\Big\} \\ &\cong X_{\mathrm{YM}}^{\ell,0}(U(n_j))_{\frac{k_j}{n_j}, \ldots, \frac{k_j}{n_j}}. \end{aligned} \tag{7.1}$$

When $k_r > 0$, define V_r by (7.1). When $k_r = 0$, define

$$V_r = \Big\{(P_r^1, Q_r^1, \ldots, P_r^\ell, Q_r^\ell) \in Sp(n_r)^{2\ell} \mid \prod_{i=1}^{\ell}[P_r^i, Q_r^i] = I_{n_r}\Big\} \cong X_{\mathrm{flat}}^{\ell,0}(Sp(n_r)).$$

Then $V_1, \dots, V_r$ are connected, and $V^{\ell,0}_{\mathrm{YM}}(Sp(n))_\mu = \prod_{j=1}^r V_j$. We have homeomorphisms

$$V^{\ell,0}_{\mathrm{YM}}(Sp(n))_\mu / Sp(n)_{X_\mu} \cong \begin{cases} \prod_{j=1}^r (V_j/U(n_j)), & k_r > 0, \\ \prod_{j=1}^{r-1}(V_j/U(n_j)) \times V_r/Sp(n_r), & k_r = 0, \end{cases}$$

and homotopy equivalences

$$V^{\ell,0}_{\mathrm{YM}}(Sp(n))_\mu{}^{hSp(n)_{X_\mu}} \sim \begin{cases} \prod_{j=1}^r V_j{}^{hU(n_j)}, & k_r > 0, \\ \prod_{j=1}^{r-1} V_j{}^{hU(n_j)} \times V_r{}^{hSp(n_r)}, & k_r = 0. \end{cases}$$

Recall that $Sp(n)$ is simply connected, so any principal $Sp(n)$-bundle over an orientable or nonorientable surface is trivial. For $i = 0, 1, 2$, let

$$\mathcal{M}(\Sigma^\ell_i, Sp(n)) = X^{\ell,i}_{\mathrm{flat}}(Sp(n))/Sp(n)$$

be the moduli space of gauge equivalence classes of flat $Sp(n)$-connections on Σ^ℓ_i.

To simplify the notation, we write

$$\mu = (\mu_1, \dots, \mu_n) = \Big(\underbrace{\frac{k_1}{n_1}, \dots, \frac{k_1}{n_1}}_{n_1}, \dots, \underbrace{\frac{k_r}{n_r}, \dots, \frac{k_r}{n_r}}_{n_r}\Big)$$

instead of

$$\mathrm{diag}\Big(\frac{k_1}{n_1} I_{n_1}, \dots, \frac{k_r}{n_r} I_{n_r}, -\frac{k_1}{n_1} I_{n_1}, \dots, -\frac{k_r}{n_r} I_{n_r}\Big).$$

Let

$$\begin{aligned} I_{Sp(n)} = \Big\{ \mu = \Big(\underbrace{\frac{k_1}{n_1}, \dots, \frac{k_1}{n_1}}_{n_1}, \dots, \underbrace{\frac{k_r}{n_r}, \dots, \frac{k_r}{n_r}}_{n_r}\Big) \Big| n_j \in \mathbb{Z}_{>0}, \\ n_1 + \cdots + n_r = n,\ k_j \in \mathbb{Z},\ \frac{k_1}{n_1} > \cdots > \frac{k_r}{n_r} \geq 0 \Big\}. \end{aligned}$$

From the discussion above, we conclude:

PROPOSITION 7.1. *Suppose that $\ell \geq 1$. Let*

$$\mu = \Big(\underbrace{\frac{k_1}{n_1}, \dots, \frac{k_1}{n_1}}_{n_1}, \dots, \underbrace{\frac{k_r}{n_r}, \dots, \frac{k_r}{n_r}}_{n_r}\Big) \in I_{Sp(n)}. \tag{7.2}$$

Then

$$X^{\ell,0}_{\mathrm{YM}}(Sp(n))_\mu / Sp(n) \cong \begin{cases} \prod_{i=1}^r \mathcal{M}(\Sigma^\ell_0, P^{n_j,k_j}), & k_r > 0, \\ \prod_{i=1}^{r-1} \mathcal{M}(\Sigma^\ell_0, P^{n_j,k_j}) \times \mathcal{M}(\Sigma^\ell_0, Sp(n_r)), & k_r = 0. \end{cases}$$

In particular, $X^{\ell,0}_{\mathrm{YM}}(Sp(n))_\mu$ is nonempty and connected. We have homotopy equivalences

$$\begin{aligned} & X^{\ell,0}_{\mathrm{YM}}(Sp(n))_\mu{}^{hSp(n)} \\ \sim & \begin{cases} \prod_{i=1}^r \Big(X^{\ell,0}_{\mathrm{YM}}(U(n_j))_{\frac{k_j}{n_j}, \dots, \frac{k_j}{n_j}} \Big)^{hU(n_j)}, & k_r > 0, \\ \prod_{i=1}^{r-1} \Big(X^{\ell,0}_{\mathrm{YM}}(U(n_j))_{\frac{k_j}{n_j}, \dots, \frac{k_j}{n_j}} \Big)^{hU(n_j)} \times X^{\ell,0}_{\mathrm{flat}}(Sp(n))^{hSp(n_r)}, & k_r = 0. \end{cases} \end{aligned}$$

PROPOSITION 7.2. *Suppose that $\ell \geq 1$. The connected components of the representation variety $X^{\ell,0}_{\mathrm{YM}}(Sp(n))$ are*

$$\{X^{\ell,0}_{\mathrm{YM}}(Sp(n))_\mu \mid \mu \in I_{Sp(n)}\}.$$

The following is an immediate consequence of Proposition 7.1.

THEOREM 7.3. *Suppose that $\ell \geq 1$, and let μ be as in (7.2). Then*

$$P_t^{Sp(n)}\left(X^{\ell,0}_{\mathrm{YM}}(Sp(n))_\mu\right)$$
$$= \begin{cases} \prod_{j=1}^r P_t^{U(n_j)}\left(X^{\ell,0}_{\mathrm{YM}}(U(n_j))_{\frac{k_j}{n_j},\dots,\frac{k_j}{n_j}}\right), & k_r > 0, \\ \prod_{j=1}^{r-1} P_t^{U(n_j)}\left(X^{\ell,0}_{\mathrm{YM}}(U(n_j))_{\frac{k_j}{n_j},\dots,\frac{k_j}{n_j}}\right) \cdot P_t^{Sp(n_r)}\left(X^{\ell,0}_{\mathrm{flat}}(Sp(n_r))\right), & k_r = 0. \end{cases}$$

7.2. Equivariant Poincaré series

Recall from Chapter 3.4.4:

$$\Delta = \{\alpha_i = \theta_i - \theta_{i+1} \mid i = 1,\dots,n-1\} \cup \{\alpha_n = 2\theta_n\}$$
$$\Delta^\vee = \{\alpha_i^\vee = e_i - e_{i+1} \mid i = 1,\dots,n-1\} \cup \{\alpha_n^\vee = e_n\}$$
$$\pi_1(H) = \bigoplus_{i=1}^n \mathbb{Z}e_i, \quad \Lambda = \bigoplus_{i=1}^{n-1} \mathbb{Z}(e_i - e_{i+1}) \oplus \mathbb{Z}e_n, \quad \pi_1(Sp(n)) = 0$$

We will apply Theorem 4.4 to the case $G_{\mathbb{R}} = Sp(n)$.

$$\varpi_{\alpha_i} = \theta_1 + \cdots + \theta_i$$

Case 1. $\alpha_n \in I$:

$$I = \{\alpha_{n_1}, \alpha_{n_1+n_2}, \dots, \alpha_{n_1+\cdots+n_{r-1}}, \alpha_n\}$$
$$L^I = GL(n_1,\mathbb{C}) \times \cdots \times GL(n_r,\mathbb{C}), \quad n_1 + \cdots + n_r = n$$
$$\dim_{\mathbb{C}} \mathfrak{z}_{L^I} - \dim_{\mathbb{C}} \mathfrak{z}_{Sp(n,\mathbb{C})} = r, \quad \dim_{\mathbb{C}} U^I = \sum_{1\leq i<j\leq r} n_i n_j + \frac{n(n+1)}{2}$$

$$\rho^I = \frac{1}{2}\sum_{i=1}^r \left(n - 2\sum_{j=1}^i n_j + n_i\right)\left(\sum_{j=1}^{n_i} \theta_{n_1+\cdots+n_{i-1}+j}\right) + \frac{n+1}{2}(\theta_1 + \cdots + \theta_n)$$

$$\langle \rho^I, \alpha^\vee_{n_1+\cdots+n_i}\rangle = \frac{n_i + n_{i+1}}{2} \text{ for } i = 1,\dots,r-1, \quad \langle \rho^I, \alpha_n^\vee\rangle = \frac{n_r+1}{2}$$

Case 2. $\alpha_n \notin I$:

$$I = \{\alpha_{n_1}, \alpha_{n_1+n_2}, \dots, \alpha_{n_1+\cdots+n_{r-1}}\}$$
$$L^I = GL(n_1,\mathbb{C}) \times \cdots \times GL(n_{r-1},\mathbb{C}) \times Sp(n_r,\mathbb{C}), \quad n_1 + \cdots + n_r = n$$
$$\dim_{\mathbb{C}} \mathfrak{z}_{L^I} - \dim_{\mathbb{C}} \mathfrak{z}_{Sp(n,\mathbb{C})} = r - 1,$$
$$\dim_{\mathbb{C}} U^I = \sum_{1\leq i<j\leq r} n_i n_j + \frac{n(n+1) - n_r(n_r+1)}{2}$$

$$\rho^I = \frac{1}{2}\sum_{i=1}^r \left(n - 2\sum_{j=1}^i n_j + n_i\right)\left(\sum_{j=1}^{n_i} \theta_{n_1+\cdots+n_{i-1}+j}\right)$$
$$+ \frac{n+1}{2}(\theta_1 + \cdots + \theta_{n_1+\cdots+n_{r-1}}) + \frac{n-n_r}{2}(\theta_{n_1+\cdots+n_{r-1}+1} + \cdots + \theta_n)$$

$$\langle \rho^I, \alpha^\vee_{n_1+\cdots+n_i}\rangle = \frac{n_i+n_{i+1}}{2} \text{ for } i=1,\ldots,r-2, \quad \langle \rho^I, \alpha^\vee_{n_1+\cdots+n_{r-1}}\rangle = \frac{n_{r-1}+1}{2}+n_r$$

Then we have the closed formula for the equivariant poinaré series for the moduli space of flat $Sp(n)$-connections:

THEOREM 7.4.

$$P_t^{Sp(n)}(X^{\ell,0}_{\mathrm{flat}}(Sp(n))) =$$

$$\sum_{r=1}^{n} \sum_{\substack{n_1,\ldots,n_r\in\mathbb{Z}_{>0}\\ \sum n_j = n}} \left((-1)^r \prod_{i=1}^{r} \frac{\prod_{j=1}^{n_i}(1+t^{2j-1})^{2\ell}}{(1-t^{2n_i})\prod_{j=1}^{n_i-1}(1-t^{2j})^2} \right.$$

$$\cdot \frac{t^{(\ell-1)(2\sum_{i<j} n_i n_j + n(n+1))}}{\left[\prod_{i=1}^{r-1}(1-t^{2(n_i+n_{i+1})})\right](1-t^{2(n_r+1)})} \cdot t^{2\sum_{i=1}^{r-1}(n_i+n_{i+1})+2(n_r+1)}$$

$$+(-1)^{r-1}\prod_{i=1}^{r-1} \frac{\prod_{j=1}^{n_i}(1+t^{2j-1})^{2\ell}}{(1-t^{2n_i})\prod_{j=1}^{n_i-1}(1-t^{2j})^2} \cdot \frac{\prod_{j=1}^{n_r}(1+t^{4j-1})^{2\ell}}{\prod_{j=1}^{2n_r}(1-t^{2j})}$$

$$\left. \cdot \frac{t^{(\ell-1)(2\sum_{i<j} n_i n_j + n(n+1) - n_r(n_r+1))}}{\left[\prod_{i=1}^{r-2}(1-t^{2(n_i+n_{i+1})})\right](1-\epsilon(r)t^{2(n_{r-1}+2n_r+1)})} t^{2\sum_{i=1}^{r-2}(n_i+n_{i+1})+2\epsilon(r)(n_{r-1}+2n_r+1)} \right)$$

where

$$\epsilon(r) = \begin{cases} 0 & r=1 \\ 1 & r>1 \end{cases}$$

EXAMPLE 7.5.

$$P_t^{Sp(1)}(X^{\ell,0}_{\mathrm{flat}}(Sp(1))) = -\frac{(1+t)^{2\ell}t^{2\ell+2}}{(1-t^2)(1-t^4)} + \frac{(1+t^3)^{2\ell}}{(1-t^2)(1-t^4)}$$

Note that $Sp(1) = SU(2) = Spin(3)$, so

$$P_t^{Sp(1)}(X^{\ell,0}_{\mathrm{flat}}(Sp(1))) = P_t^{SU(2)}(X^{\ell,0}_{\mathrm{flat}}(SU(2))) = P_t^{Spin(3)}(X^{\ell,0}_{\mathrm{flat}}(Spin(3)))$$

as expected, where $P_t^{SU(2)}(X^{\ell,0}_{\mathrm{flat}}(SU(2)))$ is calculated in Example 4.7, and that $P_t^{Spin(3)}(X^{\ell,0}_{\mathrm{flat}}(Spin(3)))$ is calculated in Example 5.7.

EXAMPLE 7.6.

$$\begin{aligned} &P_t^{Sp(2)}(X^{\ell,0}_{\mathrm{flat}}(Sp(2))) \\ = \;& -\frac{(1+t)^{2\ell}(1+t^3)^{2\ell}t^{6\ell}}{(1-t^2)^2(1-t^4)(1-t^6)} + \frac{(1+t)^{4\ell}t^{8\ell}}{(1-t^2)^2(1-t^4)^2} \\ & + \frac{(1+t^3)^{2\ell}(1+t^7)^{2\ell}}{(1-t^2)(1-t^4)(1-t^6)(1-t^8)} - \frac{(1+t)^{2\ell}(1+t^3)^{2\ell}t^{6\ell+2}}{(1-t^2)^2(1-t^4)(1-t^8)} \end{aligned}$$

Note that $Sp(2) = Spin(5)$, so

$$P_t^{Sp(2)}(X^{\ell,0}_{\mathrm{flat}}(Sp(2))) = P_t^{Spin(5)}(X^{\ell,0}_{\mathrm{flat}}(Spin(5)))$$

as expected, where $P_t^{Spin(5)}(X^{\ell,0}_{\mathrm{flat}}(Spin(5)))$ is calculated in Example 5.8.

EXAMPLE 7.7.

$$
\begin{aligned}
&P_t^{Sp(3)}(X_{\text{flat}}^{\ell,0}(Sp(3))) \\
= &-\frac{(1+t)^{2\ell}(1+t^3)^{2\ell}(1+t^5)^{2\ell}t^{12\ell-4}}{(1-t^2)^2(1-t^4)^2(1-t^6)(1-t^8)} + \frac{(1+t)^{4\ell}(1+t^3)^{2\ell}t^{16\ell-4}}{(1-t^2)^3(1-t^4)(1-t^6)^2} \\
&+\frac{(1+t)^{4\ell}(1+t^3)^{2\ell}t^{16\ell-6}}{(1-t^2)^3(1-t^4)^2(1-t^6)} - \frac{(1+t)^{6\ell}t^{18\ell-6}}{(1-t^2)^3(1-t^4)^3} \\
&+\frac{(1+t^3)^{2\ell}(1+t^7)^{2\ell}(1+t^{11})^{2\ell}}{(1-t^2)(1-t^4)(1-t^6)(1-t^8)(1-t^{10})(1-t^{12})} \\
&-\frac{(1+t)^{2\ell}(1+t^3)^{2\ell}(1+t^7)^{2\ell}t^{10\ell+2}}{(1-t^2)^2(1-t^4)(1-t^6)(1-t^8)(1-t^{12})} \\
&-\frac{(1+t)^{2\ell}(1+t^3)^{4\ell}t^{14\ell-4}}{(1-t^2)^3(1-t^4)^2(1-t^{10})} + \frac{(1+t)^{4\ell}(1+t^3)^{2\ell}t^{16\ell-4}}{(1-t^2)^3(1-t^4)^2(1-t^8)}
\end{aligned}
$$

7.3. $Sp(n)$-connections on nonorientable surfaces

We have $\overline{C}_0^{\tau} = \overline{C}_0$. Any $\mu \in \overline{C}_0^{\tau}$ is of the form

$$\mu = \operatorname{diag}(\lambda_1 I_{n_1}, \ldots, \lambda_r I_{n_r}, -\lambda_1 I_{n_1}, \ldots, -\lambda_r I_{n_r}),$$

where $\lambda_1 > \cdots > \lambda_r \geq 0$. We have

$$Sp(n)_{X_\mu} \cong \begin{cases} U(n_1) \times \cdots \times U(n_r), & \lambda_r > 0, \\ U(n_1) \times \cdots \times U(n_{r-1}) \times Sp(n_r), & \lambda_r = 0. \end{cases}$$

Suppose that $(a_1, b_1, \ldots, a_\ell, b_\ell, \epsilon c', X_\mu/2) \in X_{\text{YM}}^{\ell,1}(Sp(n))$, where

$$\epsilon = \begin{pmatrix} 0 & -I_n \\ I_n & 0 \end{pmatrix} \in Sp(n)$$

is defined as in Example 4.12. Notice that here $\epsilon^2 \neq 1$. Then

$$\exp(X_\mu/2)\epsilon c' \epsilon c' = \prod_{i=1}^{\ell} [a_i, b_i]$$

where a_i, b_i, $c' \in Sp(n)_{X_\mu}$. Note that $\epsilon c' \epsilon c' = -\bar{c'} c'$ where $\bar{c'}$ is the complex conjugate of c', so

$$\exp(X_\mu/2)(-\bar{c'}c') \in (Sp(n)_{X_\mu})_{ss} \cong \begin{cases} SU(n_1) \times \cdots \times SU(n_r), & \lambda_r > 0, \\ SU(n_1) \times \cdots \times SU(n_{r-1}) \times Sp(n_r), & \lambda_r = 0. \end{cases}$$

In order that this is nonempty, we need $1 = \det(-e^{\pi\sqrt{-1}\lambda_j} I_{n_j})$, i.e.,

$$\lambda_j = \frac{2k_j}{n_j} - 1, \quad k_j \in \mathbb{Z}, \quad j = 1, \ldots, r.$$

Similarly, suppose that $(a_1, b_1, \ldots, a_\ell, b_\ell, d, \epsilon c', X_\mu/2) \in X_{\text{YM}}^{\ell,2}(Sp(n))$. Then

$$\exp(X_\mu/2)(\epsilon c')d(\epsilon c')^{-1}d = \prod_{i=1}^{\ell} [a_i, b_i],$$

or equivalently,

$$\exp(X_\mu/2)(-\bar{c'}\bar{d}\bar{c'}^{-1}d) \in (Sp(n)_{X_\mu})_{ss} \cong \begin{cases} SU(n_1)\times\cdots\times SU(n_r), & \lambda_r>0,\\ SU(n_1)\times\cdots\times SU(n_{r-1})\times Sp(n_r), & \lambda_r=0.\end{cases}$$

Again, we need

$$\lambda_j = \frac{2k_j}{n_j}-1,\quad k_j\in\mathbb{Z},\quad j=1,\dots,r.$$

We conclude that for nonorientable surfaces, either

$$\mu = \mathrm{diag}\Big((\frac{2k_1}{n_1}-1)I_{n_1},\dots,(\frac{2k_r}{n_r}-1)I_{n_r},-(\frac{2k_1}{n_1}-1)I_{n_1},\dots,-(\frac{2k_r}{n_r}-1)I_{n_r}\Big),$$

where

$$k_j\in\mathbb{Z},\quad \frac{k_1}{n_1}>\cdots>\frac{k_r}{n_r}>\frac{1}{2},$$

or

$$\begin{aligned}\mu = \mathrm{diag}\Big(&(\frac{2k_1}{n_1}-1)I_{n_1},\dots,(\frac{2k_{r-1}}{n_{r-1}}-1)I_{n_{r-1}},0I_{n_r},\\ &-(\frac{2k_1}{n_1}-1)I_{n_1},\dots,-(\frac{2k_{r-1}}{n_{r-1}}-1)I_{n_{r-1}},0I_{n_r}\Big),\end{aligned}$$

where

$$k_j\in\mathbb{Z},\quad \frac{k_1}{n_1}>\cdots>\frac{k_{r-1}}{n_{r-1}}>\frac{1}{2}.$$

Recall the for each μ, the ϵ-reduced representation varieties are

$$\begin{aligned} V^{\ell,1}_{\mathrm{YM}}(Sp(n))_\mu &= \{(a_1,b_1,\dots,a_\ell,b_\ell,c')\in Sp(n)^{2\ell+1}_{X_\mu}\mid \\ &\qquad \prod_{i=1}^{\ell}[a_i,b_i]=\exp(X_\mu/2)\epsilon c'\epsilon c'\},\\ V^{\ell,2}_{\mathrm{YM}}(Sp(n))_\mu &= \{(a_1,b_1,\dots,a_\ell,b_\ell,d,c')\in Sp(n)^{2\ell+2}_{X_\mu}\mid \\ &\qquad \prod_{i=1}^{\ell}[a_i,b_i]=\exp(X_\mu/2)\epsilon c'd(\epsilon c')^{-1}d\}.\end{aligned}$$

Let $i=1,\cdots,\ell$. When $\lambda_r>0$, write

$$\begin{aligned} a_i &= \mathrm{diag}\left(A^i_1,\dots,A^i_r,\bar{A}^i_1,\dots,\bar{A}^i_r\right), & b_i &= \mathrm{diag}\left(B^i_1,\dots,B^i_r,\bar{B}^i_1,\dots,\bar{B}^i_r\right),\\ c' &= \mathrm{diag}\left(C_1,\dots,C_r,\bar{C}_1,\dots,\bar{C}_r\right), & d &= \mathrm{diag}\left(D_1,\dots,D_r,\bar{D}_1,\dots,\bar{D}_r\right),\end{aligned}$$

where $A^i_j, B^i_j, C_j, B_j \in U(n_j)$. When $\lambda_r=0$, write

$$a_i = \begin{pmatrix} A^i & & & 0\\ & A^i_r & & -\bar{E}^i_r\\ & & \bar{A}^i & \\ 0 & E^i_r & & \bar{A}^i_r\end{pmatrix},\quad b_i = \begin{pmatrix} B^i & & & 0\\ & B^i_r & & -\bar{F}^i_r\\ & & \bar{A}^i & \\ 0 & F^i_r & & \bar{B}^i_r\end{pmatrix},$$

$$c' = \begin{pmatrix} C & & & 0\\ & C_r & & -\bar{H}_r\\ & & \bar{C} & \\ 0 & H_r & & \bar{C}_r\end{pmatrix},\quad d = \begin{pmatrix} D & & & 0\\ & D_r & & -\bar{G}_r\\ & & \bar{D} & \\ 0 & G_r & & \bar{D}_r\end{pmatrix},$$

where

$$
\begin{aligned}
&A^i = \operatorname{diag}\left(A^i_1, \ldots, A^i_{r-1}\right), \quad B^i = \operatorname{diag}\left(B^i_1, \ldots, B^i_{r-1}\right), \\
&C = \operatorname{diag}\left(C_1, \ldots, C_{r-1}\right), \quad D = \operatorname{diag}\left(D_1, \ldots, D_{r-1}\right), \\
&A^i_j, B^i_j, C_j, D_j \in U(n_j), \quad j = 1, \ldots, r-1, \\
&P^i = \begin{pmatrix} A^i_r & -\bar{E}^i_r \\ E^i_r & \bar{A}^i_r \end{pmatrix}, Q^i = \begin{pmatrix} B^i_r & -\bar{F}^i_r \\ F^i_r & \bar{B}^i_r \end{pmatrix} \in Sp(n_r) \subset U(2n_r), \\
&S^i = \begin{pmatrix} C_r & -\bar{H}_r \\ H_r & \bar{C}_r \end{pmatrix}, R^i = \begin{pmatrix} D_r & -\bar{G}_r \\ G_r & \bar{D}_r \end{pmatrix} \in Sp(n_r) \subset U(2n_r).
\end{aligned}
$$

Let $\epsilon = \begin{pmatrix} 0 & -I_{n_r} \\ I_{n_r} & 0 \end{pmatrix} \in Sp(n_r)$. For $j = 1, \ldots, r-1$, define

(7.3)

$$
\begin{aligned}
V_j &= \Big\{(A^1_j, B^1_j, \ldots, A^\ell_j, B^\ell_j, C_j) \in U(n_j)^{2\ell+1} \mid \prod_{i=1}^{\ell}[A^i_j, B^i_j] = e^{-2\pi\sqrt{-1}\frac{k_j}{n_j}}(\overline{C}_j C_j)\Big\} \\
&\cong \tilde{V}^{\ell,1}_{n_j,k_j},
\end{aligned}
$$

where $\tilde{V}^{\ell,1}_{n_j,k_j}$ is the twisted representation variety defined in (4.7) of Section 4.6. $\tilde{V}^{\ell,1}_{n_j,k_j}$ is nonempty if $\ell \geq 1$. We have shown that $\tilde{V}^{\ell,1}_{n_j,k_j}$ is connected if $\ell \geq 2$ (Proposition 4.13). When $\lambda_r > 0$, define V_r by (7.3). When $\lambda_r = 0$, define

$$
\begin{aligned}
V_r &= \Big\{(P^1_r, Q^1_r, \ldots, P^\ell_r, Q^\ell_r, S_r) \in Sp(n_r)^{2\ell+1} \mid \prod_{i=1}^{\ell}[P^i_r, Q^i_r] = (\epsilon S_r)^2\Big\} \\
&\overset{(S'_r=\epsilon S_r)}{\cong} \Big\{(P^1_r, Q^1_r, \ldots, P^\ell_r, Q^\ell_r, S'_r) \in Sp(n_r)^{2\ell+1} \mid \prod_{i=1}^{\ell}[P^i_r, Q^i_r] = (S'_r)^2\Big\} \\
&\cong X^{\ell,1}_{\mathrm{flat}}(Sp(n_0)).
\end{aligned}
$$

Then $V^{\ell,1}_{\mathrm{YM}}(Sp(n))_\mu = \prod_{j=1}^r V_j$.

Similarly, for $j = 1, \ldots, r-1$, define

(7.4)

$$
\begin{aligned}
V_j = \Big\{&(A^1_j, B^1_j, \ldots, A^\ell_j, B^\ell_j, D_j, C_j) \in (U(n_j))^{2\ell+2} \mid \\
&\prod_{i=1}^{\ell}[A^i_j, B^i_j] = \exp^{-2\pi\sqrt{-1}\frac{k_j}{n_j}} I_{n_j}\bar{C}_j\bar{D}_j\bar{C}_j^{-1}D_j\Big\} \quad \cong \quad \tilde{V}^{\ell,2}_{n_j,k_j},
\end{aligned}
$$

where $\tilde{V}^{\ell,2}_{n_j,k_j}$ is the twisted representation variety defined in (4.8) of Section 4.6. $\tilde{V}^{\ell,2}_{n_j,k_j}$ is nonempty if $\ell \geq 1$. We have shown that $\tilde{V}^{\ell,2}_{n_j,k_j}$ is connected if $\ell \geq 4$ (Proposition 4.13). When $\lambda_r > 0$, define V_r by (7.4). When $\lambda_r = 0$, define

$$
\begin{aligned}
V_r &= \Big\{(P^1_r, Q^1_r, \ldots, P^\ell_r, Q^\ell_r, R_r, S_r) \in Sp(n_r)^{2\ell+2} \mid \prod_{i=1}^{\ell}[P^i_r, Q^i_r] = \epsilon S_r R_r(\epsilon S_r)^{-1}R_r\Big\} \\
&\overset{(S'_r=\epsilon S_r)}{\cong} \Big\{(P^1_r, Q^1_r, \ldots, P^\ell_r, Q^\ell_r, R_r, S'_r) \in Sp(n_r)^{2\ell+2} \mid \prod_{i=1}^{\ell}[P^i_r, Q^i_r] = S'_r R_r(S'_r)^{-1}R_r\Big\} \\
&\cong X^{\ell,2}_{\mathrm{flat}}(Sp(n_r)).
\end{aligned}
$$

Then $V^{\ell,2}_{\mathrm{YM}}(Sp(n))_\mu = \prod_{j=1}^r V_j$.

Let $U(n_j)$ act on $V_j = \tilde{V}^{\ell,i}_{n_j,k_j}$ by (4.9) and (4.10) in Section 4.6 when $i=1$ and when $i=2$, respectively. Then we have homeomorphisms

$$V^{\ell,i}_{\mathrm{YM}}(Sp(n))_\mu / Sp(n)_{X_\mu} \cong \begin{cases} \prod_{j=1}^{r}(V_j/U(n_j)), & \lambda_r > 0, \\ \prod_{j=1}^{r-1}(V_j/U(n_j)) \times V_r/Sp(n_r), & \lambda_r = 0, \end{cases}$$

and homotopy equivalences

$$V^{\ell,i}_{\mathrm{YM}}(Sp(n))_\mu{}^{Sp(n)_{X_\mu}} \sim \begin{cases} \prod_{j=1}^{r} V_j{}^{hU(n_j)}, & \lambda_r > 0, \\ \prod_{j=1}^{r-1} V_j{}^{hU(n_j)} \times V_r{}^{hSp(n_r)}, & \lambda_r = 0. \end{cases}$$

To simplify the notation, we write

$$\mu = (\mu_1, \dots, \mu_n) = \Big(\underbrace{\frac{2k_1}{n_1} - 1, \dots, \frac{2k_1}{n_1} - 1}_{n_1}, \dots, \underbrace{\frac{2k_r}{n_r} - 1, \dots, \frac{2k_r}{n_r} - 1}_{n_r}\Big) \tag{7.5}$$

instead of

$$\mathrm{diag}\Big(\big(\frac{2k_1}{n_1} - 1\big)I_{n_1}, \dots, \big(\frac{2k_r}{n_r} - 1\big)I_{n_r}, -\big(\frac{2k_1}{n_1} - 1\big)I_{n_1}, \dots, -\big(\frac{2k_r}{n_r} - 1\big)\big)I_{n_r}\Big),$$

and write

(7.6)

$$\mu = (\mu_1, \dots, \mu_n) = \Big(\underbrace{\frac{2k_1}{n_1} - 1, \dots, \frac{2k_1}{n_1} - 1}_{n_1}, \dots, \underbrace{\frac{2k_{r-1}}{n_{r-1}} - 1, \dots, \frac{2k_{r-1}}{n_{r-1}} - 1}_{n_{r-1}}, \underbrace{0, \dots, 0}_{n_r}\Big)$$

instead of

$$\mathrm{diag}\Big(\big(\frac{2k_1}{n_1} - 1\big)I_{n_1}, \dots, \big(\frac{2k_{r-1}}{n_{r-1}} - 1\big)I_{n_{r-1}}, 0I_{n_r},$$
$$-\big(\frac{2k_1}{n_1} - 1\big)I_{n_1}, \dots, -\big(\frac{2k_{r-1}}{n_{r-1}} - 1\big)\big)I_{n_{r-1}}, 0I_{n_r}\Big).$$

Let

$$\begin{aligned} \hat{I}_{Sp(n)} &= \Big\{\mu = \Big(\underbrace{\frac{2k_1}{n_1} - 1, \dots, \frac{2k_1}{n_1} - 1}_{n_1}, \dots, \underbrace{\frac{2k_r}{n_r} - 1, \dots, \frac{2k_r}{n_r} - 1}_{n_r}\Big) \mid n_j \in \mathbb{Z}_{>0}, \\ &\qquad n_1 + \cdots + n_r = n,\ k_j \in \mathbb{Z},\ \frac{k_1}{n_1} > \cdots > \frac{k_r}{n_r} > \frac{1}{2}\Big\} \\ &\cup \Big\{\mu = \Big(\underbrace{\frac{2k_1}{n_1} - 1, \dots, \frac{2k_1}{n_1} - 1}_{n_1}, \dots, \underbrace{\frac{2k_{r-1}}{n_{r-1}} - 1, \dots, \frac{2k_{r-1}}{n_{r-1}} - 1}_{n_{r-1}}, \underbrace{0, \dots, 0}_{n_r}\Big) \mid \\ &\qquad n_j \in \mathbb{Z}_{>0},\ n_1 + \cdots + n_r = n,\ k_j \in \mathbb{Z},\ \frac{k_1}{n_1} > \cdots > \frac{k_{r-1}}{n_{r-1}} > \frac{1}{2}\Big\} \end{aligned}$$

PROPOSITION 7.8. *Suppose that $\ell \geq 2i$, where $i = 1, 2$, and let $\mu \in \hat{I}_{Sp(n)}$.*

(i) *If μ is of the form (7.5), then*

$$X^{\ell,i}_{\mathrm{YM}}(Sp(n))_\mu / Sp(n) \cong \prod_{j=1}^{r} \tilde{\mathcal{M}}^{\ell,i}_{n_j,k_j}.$$

We have a homotopy equivalence

$$X^{\ell,i}_{\mathrm{YM}}(Sp(n))_\mu{}^{hSp(n)} \sim \prod_{j=1}^{r} (\tilde{V}^{\ell,i}_{n_j,k_j})^{hU(n_j)}.$$

(ii) *If μ is of the form (7.6), then*

$$X^{\ell,i}_{\mathrm{YM}}(Sp(n))_\mu / Sp(n) \cong \prod_{j=1}^{r-1} \tilde{\mathcal{M}}^{\ell,i}_{n_j,k_j} \times \mathcal{M}(\Sigma^\ell_i, Sp(n_r)).$$

We have a homotopy equivalence

$$X^{\ell,i}_{\mathrm{YM}}(Sp(n))_\mu{}^{hSp(n)} \sim \prod_{j=1}^{r-1} (\tilde{V}^{\ell,i}_{n_j,k_j})^{hU(n_j)} \times X^{\ell,i}_{\mathrm{flat}}(Sp(n_r))^{hSp(n_r)}.$$

In particular, $X^{\ell,i}_{\mathrm{YM}}(Sp(n))_\mu$ is nonempty and connected.

PROPOSITION 7.9. *Suppose that $\ell \geq 2i$, where $i = 1, 2$. The connected components of $X^{\ell,i}_{\mathrm{YM}}(Sp(n))$ are*

$$\{X^{\ell,i}_{\mathrm{YM}}(Sp(n))_\mu \mid \mu \in \hat{I}_{Sp(n)}\}.$$

Notice that, the set $\{\mu = \mathrm{diag}(\mu_1, \ldots, \mu_n, -\mu_1, \ldots, -\mu_n) | (\mu_1, \ldots, \mu_n) \in \hat{I}_{Sp(n)}\}$ is a *proper* subset of $\{\mu \in (\Xi^I_+)^\tau | I \subseteq \Delta, \tau(I) = I\}$ as mentioned in Section 4.5.

The following is an immediate consequence of Proposition 7.8.

THEOREM 7.10. *Suppose that $\ell \geq 2i$, where $i = 1, 2$, and let $\mu \in \hat{I}_{Sp(n)}$.*

(i) *If μ is of the form (7.5), then*

$$P^{Sp(n)}_t\left(X^{\ell,i}_{\mathrm{YM}}(Sp(n))_\mu\right) = \prod_{j=1}^{r} P^{U(n_j)}_t(\tilde{V}^{\ell,i}_{n_j,k_j}).$$

(ii) *If μ is of the form (7.6), then*

$$P^{Sp(n)}_t\left(X^{\ell,i}_{\mathrm{YM}}(Sp(n))_\mu\right) = \prod_{j=1}^{r-1} P^{U(n_j)}_t(\tilde{V}^{\ell,i}_{n_j,k_j}) \cdot P^{Sp(n_r)}_t\left(X^{\ell,i}_{\mathrm{flat}}(Sp(n_r))\right).$$

APPENDIX A

Remarks on Laumon-Rapoport Formula

In this appendix, we explain how to use the argument in [**LR**] to obtain Theorem 4.4, which is a slightly modified version of [**LR**, Theorem 3.4]. We work over $\mathbb{C}$.

A.1. Notation

The following is a correspondence between the notation in [**FM**] (which we followed closely in Chapter 3) and that in [**LR**].

	[**LR**]	[**FM**]
minimal parabolic subgroup (Borel)	P_0	
Cartan of G	M_0	H
parabolic subgroup	$P = M_P N_P$	$P = LU$
Levi subgroup	M_P	L
unipotent radical	N_P	U
center of the Levi subgroup	Z_P	$Z(L)$
connected center of M_P	A_P	$Z(L)_0$
	$A'_P \subset M_{P,\mathrm{ab}}$	$L/[L,L] = Z(L)_0/Z(L)_0 \cap [L,L]$
	$X_*(A_P)$	$\pi_1(Z(L)_0)$
	$X_*(A'_P)$	$\pi_1(H)/\hat{\Lambda}_L = \pi_1(L/[L,L])$
	$X_*(A'_{P_0})$	$\pi_1(H)$
	$\mathfrak{a}_0 = \mathfrak{a}_{P_0}$	$\mathfrak{h}_{\mathbb{R}}$
	$\mathfrak{a}_P = \mathbb{R} \otimes X_*(A_P)$ $= \mathbb{R} \otimes X_*(A'_P)$	$(\mathfrak{z}_L)_{\mathbb{R}}$
	$\mathfrak{a}_G = \mathbb{R} \otimes X_*(A_G)$ $= \mathbb{R} \otimes X_*(A'_G)$	$(\mathfrak{z}_G)_{\mathbb{R}} \cong \mathfrak{h}_{\mathbb{R}}/V^*$
	$\mathfrak{a}_0^G = \mathfrak{a}_{P_0}^G \subset \mathfrak{a}_0$	$V^* = \Lambda \otimes \mathbb{R} \subset \mathfrak{h}_{\mathbb{R}}$
root system	$\Phi_0 = \Phi_{P_0} \subset \mathfrak{a}_0^\vee$	$R \subset \mathfrak{h}_{\mathbb{R}}^*$
set of positive roots	$\Phi_0^+ = \Phi_{P_0}^+ \subset \Phi_0$	$R^+ \subset R$
set of simple roots	$\Delta_0 = \Delta_{P_0} \subset \Phi_0^+$	$\Delta \subset R^+$
coroot lattice of G	$\bigoplus_{\alpha \in \Delta_0} \mathbb{Z}\alpha^\vee$	$\Lambda = \bigoplus_{\alpha \in \Delta} \mathbb{Z}\alpha^\vee \subset \pi_1(H)$

In this appendix, we will closely follow the notation in [**LR**]. We will not repeat most of the definitions in [**LR**].

Following [**LR**], if $P \subset Q \subset R$ are three parabolic subgroups of G, there are canonical splittings $\mathfrak{a}_P = \mathfrak{a}_P^Q \oplus \mathfrak{a}_Q^R \oplus \mathfrak{a}_R$ and $\mathfrak{a}_P^* = \mathfrak{a}_P^{Q*} \oplus \mathfrak{a}_Q^{R*} \oplus \mathfrak{a}_R^*$. Given $H \in \mathfrak{a}_P$, we denote by $[H]^Q$, $[H]^R_Q$, and $[H]_R$ the canonical projections of H onto $\mathfrak{a}^Q_P$, $\mathfrak{a}^R_Q$, and $\mathfrak{a}_R$, respectively. The components of $\beta \in \mathfrak{a}^*_P$ in $\mathfrak{a}^{Q*}_P$, $\mathfrak{a}^{R*}_Q$, and $\mathfrak{a}^*_R$ are $\beta|_{\mathfrak{a}^Q_P}$, $\beta|_{\mathfrak{a}^R_Q}$, and $\beta|_{\mathfrak{a}_R}$, respectively. Given $\alpha \in \Delta_P = \Delta^G_P \subset \mathfrak{a}^{G*}_P$, let $\tilde{\alpha}$ denote the unique element in $\Delta_0 \subset \mathfrak{a}^{G*}_{P_0}$ such that $\tilde{\alpha}|_{\mathfrak{a}^G_P} = \alpha$. Then $\tilde{\alpha}^\vee \in \mathfrak{a}^G_{P_0}$ and $\alpha^\vee = [\tilde{\alpha}^\vee]_P \in \mathfrak{a}^G_P$. The subset I^P of the set of simple roots in **[FM]** corresponds to $\Delta_P = \Delta^G_P$ in the following way:

$$
\begin{aligned}
I^P &= \{\tilde{\alpha} \mid \alpha \in \Delta_P\} \subset \Delta_0 \subset \mathfrak{a}^{G*}_{P_0} \\
\Delta_P &= \{\beta|_{\mathfrak{a}_P} \mid \beta \in I^P\} \subset \mathfrak{a}^{G*}_P \\
\Delta^P_{P_0} &= \{\beta|_{\mathfrak{a}^P_{P_0}} \mid \beta \in \Delta_0 \setminus I^P\} \subset \mathfrak{a}^{P*}_{P_0} \\
\Delta^Q_P &= \{\beta|_{\mathfrak{a}^Q_P} \mid \beta \in I^P \setminus I^Q\} \subset \mathfrak{a}^{Q*}_P
\end{aligned}
$$

We continue the table of correspondence between notations in **[LR]** and **[FM]**:

	[LM]	**[FM]**
	$X_*(A'_P)$	$\pi_1(H)/\widehat{\Lambda}_L$
	$\Lambda^G_P = X_*(A'_P)/\bigoplus_{\alpha\in\Delta^G_P}\mathbb{Z}\alpha^\vee$	$\pi_1(H)/(\widehat{\Lambda}_L \oplus \bigoplus_{\alpha\in I^P}\mathbb{Z}\alpha^\vee)$
Topological type of G-bundle	$\Lambda^G_{P_0}$	$\pi_1(H)/\Lambda = \pi_1(G)$
Topological type of M_P-bundle	$\Lambda^P_{P_0}$	$\pi_1(H)/\Lambda_L = \pi_1(L)$

Given a parabolic subgroup P of G, the topological type of an M_P-bundle is given by $\nu_P \in \Lambda^P_{P_0} \cong \pi_1(M_P)$. The slope of an M_P-bundle is given by $\nu'_P \in X_*(A'_P)$. The commutative diagram in Section 3.3 can be rewritten as follows:

$$
\begin{array}{ccccccc}
 & & 0 & & 0 & & \\
 & & \downarrow & & \downarrow & & \\
0 & \longrightarrow & \widehat{\Lambda}_P/\Lambda_P & \xrightarrow{j_{ss}} & \widehat{\Lambda}/\Lambda & \xrightarrow{\oplus_{\alpha\in\Delta^G_P}\varpi_{\tilde{\alpha}}} & \oplus_{\alpha\in\Delta^G_P}\mathbb{Q}/\mathbb{Z} \\
 & & \downarrow{\scriptstyle i_P} & & \downarrow{\scriptstyle i_G} & & \| \\
 & & \Lambda^P_{P_0} & \xrightarrow{[\cdot]_G} & \Lambda^G_{P_0} & \xrightarrow{\oplus_{\alpha\in\Delta^G_P}\varpi_{\tilde{\alpha}}} & \oplus_{\alpha\in\Delta^G_P}\mathbb{Q}/\mathbb{Z} \\
 & & \downarrow{\scriptstyle p_P} & & \downarrow{\scriptstyle p_G} & & \\
 & & X_*(A'_P) & \xrightarrow{[\cdot]_G} & X_*(A'_G) & & \\
 & & \downarrow & & \downarrow & & \\
 & & 0 & & 0 & &
\end{array}
$$

Here $\Lambda_P = \oplus_{\alpha\in\Delta^P_{P_0}}\mathbb{Z}\alpha^\vee \subset X_*(A'_{P_0})$, and $\widehat{\Lambda}_P$ is the saturation of Λ_P in $X_*(A'_{P_0})$. Let ν'_P and ν'_G denote the projections $p_P(\nu_P)$ and $p_G(\nu_G)$, respectively.

Recall that $\{\varpi_\alpha \mid \alpha \in \Delta_0\}$ is a basis of the real vector space $\mathfrak{a}^{G*}_{P_0}$ which is dual to the basis $\{\alpha^\vee \mid \alpha \in \Delta_0\}$ of $\mathfrak{a}^G_{P_0}$. Given $\alpha \in \Delta_0$, we extend $\varpi_\alpha : \mathfrak{a}^G_{P_0} \to \mathbb{R}$ to $\varpi_\alpha : \mathfrak{a}_0 = \mathfrak{a}^G_{P_0} \oplus \mathfrak{a}_G \to \mathbb{R}$ by zero on $\mathfrak{a}_G$. Then ϖ_α takes integral values on $\oplus_{\alpha\in\Delta_0}\mathbb{Z}\alpha^\vee \subset \mathfrak{a}^G_{P_0} \subset \mathfrak{a}_0$, and takes rational values on $X_*(A'_{P_0}) \subset \mathfrak{a}_0$. So it induces a map

$$\varpi_\alpha : \Lambda_{P_0}^Q = X_*(A'_{P_0}) \Big/ \bigoplus_{\alpha\in\Delta_{P_0}^Q} \mathbb{Z}\alpha^\vee \to \mathbb{Q}/\mathbb{Z}$$

where Q is any parabolic subgroup of G. More explicitly, given $\nu_Q \in \Lambda_{P_0}^Q$, let $X \in X_*(A'_{P_0})$ be a representative of ν_Q. Then $\varpi_\alpha(\nu_Q) = \varpi_\alpha(X) + \mathbb{Z}$.

A.2. Inversion formulas

Let A be a fixed topological abelian group. In [**LR**], Laumon and Rapoport introduced the notion of $\widehat{\Gamma}$-converging functions and Γ-converging functions from $\mathfrak{P}$ to A, where

$$\mathfrak{P} = \{(P, \nu'_P) \mid P \in \mathcal{P}, \nu'_P \in X_*(A'_P)\}.$$

We will introduce similar notion for functions from $\mathfrak{T}$ to A, where

$$\mathfrak{T} = \{(P, \nu_P) \mid P \in \mathcal{P}, \nu_P \in \Lambda_{P_0}^P\}.$$

DEFINITION A.1. Let $\mathfrak{T} = \{(P, \nu_P) \mid P \in \mathcal{P}, \nu_P \in \Lambda_{P_0}^P\}$, and let A be a fixed topological abelian group. A function $a : \mathfrak{T} \to A$ is $\widehat{\Gamma}$*-converging* if for each standard parabolic subgroup $P \subset Q$ of G and each $\nu_Q \in \Lambda_{P_0}^Q$, the finite sum

$$\sum_{\substack{\nu_P\in\Lambda_{P_0}^P\\ [\nu_P]_Q=\nu_Q}} \widehat{\Gamma}_P^Q([\nu'_P]^Q, T)a(P, \nu_P)$$

admits a limit as $T \in \mathfrak{a}_P^{Q+}$ goes to infinity. If this is the case, we shall denote this limit by

$$\sum_{\substack{\nu_P\in\Lambda_{P_0}^P\\ [\nu_P]_Q=\nu_Q}} \widehat{\tau}_P^Q([\nu'_P]^Q)a(P, \nu_P) .$$

A function $b : \mathfrak{T} \to A$ is Γ*-converging* if for each standard parabolic subgroup $P \subset Q$ of G and each $\nu_Q \in \Gamma_{P_0}^Q$, the finite sum

$$\sum_{\substack{\nu_P\in\Lambda_{P_0}^P\\ [\nu_P]_Q=\nu_Q}} \Gamma_P^Q([\nu'_P]^Q, T)b(P, \nu_P)$$

admits a limit as $T \in \mathfrak{a}_P^{Q+}$ goes to infinity. If this is the case, we shall denote this limit by

$$\sum_{\substack{\nu_P\in\Lambda_{P_0}^P\\ [\nu_P]_Q=\nu_Q}} \tau_P^Q([\nu'_P]^Q)b(P, \nu_P) .$$

The following inversion formula is an analogue of [**LR**, Theorem 2.1].

THEOREM A.2. *For each* $\widehat{\Gamma}$*-converging function* $a : \mathfrak{T} \to A$*, there exists a unique* Γ*-converging function* $b : \mathfrak{T} \to A$ *such that, for each* $(Q, \nu_Q) \in \mathfrak{T}$*, we have*

$$a(Q, \nu_Q) = \sum_{\substack{P\in\mathcal{P}\\ P\subset Q}} \sum_{\substack{\nu_P\in\Lambda_{P_0}^P\\ [\nu_P]_Q=\nu_Q}} \tau_P^Q([\nu'_P]^Q)b(P, \nu_P) .$$

The function b is given by the following formula : for each $(Q,\nu_Q)\in\mathfrak{T}$, *we have*

$$b(Q,\nu_Q)=\sum_{\substack{P\in\mathcal{P}\\ P\subset Q}}(-1)^{\dim(\mathfrak{a}_P^Q)}\sum_{\substack{\nu_P\in\Lambda_{P_0}^P\\ [\nu_P]_Q=\nu_Q}}\widehat{\tau}_P^Q([\nu_P']^Q)a(P,\nu_P)\ .$$

Theorem A.2 is an easy consequence of the following two lemmas:

LEMMA A.3 (Langlands). *For any standard parabolic subgroups* $P\subset R$ *of* G *and any* $H\in\mathfrak{a}_P^R$, *we have*

$$\sum_{P\subset Q\subset R}(-1)^{\dim(\mathfrak{a}_Q^R)}\tau_P^Q([H]^Q)\widehat{\tau}_Q^R([H]_Q)=\delta_P^R \tag{A.1}$$

and

$$\sum_{P\subset Q\subset R}(-1)^{\dim(\mathfrak{a}_P^Q)}\widehat{\tau}_P^Q([H]^Q)\tau_Q^R([H]_Q)=\delta_P^R. \tag{A.2}$$

LEMMA A.4 (Arthur). *If* $T\in\mathfrak{a}_P^{R+}\subset{}^+\mathfrak{a}_P^R$, *the function* $H\mapsto\Gamma_P^R(H,T)$ *(resp.* $H\mapsto\hat{\Gamma}_P^R(H,T)$*) is the characteristic function of the bounded subset*

$$\{H\in\mathfrak{a}_P^R\mid\langle\alpha,H\rangle>0,\langle\varpi_\alpha,H\rangle\le\langle\varpi_\alpha,T\rangle,\forall\alpha\in\Delta_P^R\}\subset\mathfrak{a}_P^{R+}$$

(resp.

$$\{H\in\mathfrak{a}_P^R\mid\langle\varpi_\alpha^R,H\rangle>0,\langle\alpha,H\rangle\le\langle\alpha,T\rangle,\forall\alpha\in\Delta_P^R\}\subset{}^+\mathfrak{a}_P^R\,)$$

of $\mathfrak{a}_P^R$.

PROOF OF THEOREM A.2.

$$\begin{aligned}
&\sum_{\substack{Q\in\mathcal{P}\\ Q\subset R}}\sum_{\substack{\nu_Q\in\Lambda_{P_0}^Q\\ [\nu_Q]_R=\nu_R}}\tau_Q^R([\nu_Q']^R)b(Q,\nu_Q)\\
=&\sum_{\substack{Q\in\mathcal{P}\\ Q\subset R}}\sum_{\substack{\nu_Q\in\Lambda_{P_0}^Q\\ [\nu_Q]_R=\nu_R}}\tau_Q^R([\nu_Q']^R)\sum_{\substack{P\in\mathcal{P}\\ P\subset Q}}(-1)^{\dim(\mathfrak{a}_P^Q)}\sum_{\substack{\nu_P\in\Lambda_{P_0}^P\\ [\nu_P]_Q=\nu_Q}}\widehat{\tau}_P^Q([\nu_P']^Q)a(P,\nu_P)\\
=&\sum_{P\subset Q\subset R}\sum_{\substack{\nu_P\in\Lambda_{P_0}^P\\ [\nu_P]_R=\nu_R}}(-1)^{\dim(\mathfrak{a}_P^Q)}\widehat{\tau}_P^Q([\nu_P']^Q)\tau_Q^R([\nu_P']_Q^R)a(P,\nu_P)
\end{aligned}$$

For fixed ν_P, we have

$$\begin{aligned}
&\sum_{P\subset Q\subset R}(-1)^{\dim(\mathfrak{a}_P^Q)}\widehat{\tau}_P^Q([\nu_P']^Q)\tau_Q^R([\nu_P']_Q^R)=\\
&\qquad\sum_{P\subset Q\subset R}(-1)^{\dim(\mathfrak{a}_P^Q)}\widehat{\tau}_P^Q([[\nu_P']^R]^Q)\tau_Q^R([[\nu_P']^R]_Q)=\delta_P^R
\end{aligned}$$

where the last equality follows from (A.2) in Lemma A.3. So

$$\sum_{\substack{Q\in\mathcal{P}\\ Q\subset R}}\sum_{\substack{\nu_Q\in\Lambda_{P_0}^Q\\ [\nu_Q]_R=\nu_R}}\tau_Q^R([\nu_Q']^R)b(Q,\nu_Q)=\sum_{\substack{P\in\mathcal{P}\\ P\subset R}}\sum_{\substack{\nu_P\in\Lambda_{P_0}^P\\ [\nu_P]_R=\nu_R}}\delta_P^Ra(P,\nu_P)=a(R,\nu_R)$$

□

Now we consider a special case of Theorem A.2. For any $P \in \mathcal{P}$, fix $n_P \in \mathbb{Z}_{\geq 0}$ and $\epsilon_0^P \in \mathfrak{a}_0^{P*} \subset \mathfrak{a}_0^*$ such that for any standard parabolic subgroups $P \subset Q$ of G,

$$n_P \geq n_Q, \quad (\epsilon_0^Q - \epsilon_0^P)\big|_{\mathfrak{a}_0^P} = 0, \quad \langle \epsilon_P^Q, \alpha^\vee \rangle \in \mathbb{Z}_{>0} \quad \forall \alpha \in \Delta_P^Q,$$

where $\epsilon_P^Q = (\epsilon_0^Q - \epsilon_0^P)\big|_{\mathfrak{a}_P^Q}$. (Here we use ϵ_P^Q instead of δ_P^Q, which is used in [**LR**], to avoid confusion with the δ_P^R in Lemma A.3.)

We have the following analogue of [**LR**, Lemma 2.3]:

LEMMA A.5. *For each $(Q, \nu_Q) \in \mathfrak{T}$ and each standard parabolic subgroup $P \subset Q$ of G, we have*

$$\sum_{\substack{\nu_P \in \Lambda_{P_0}^P \\ [\nu_P]_Q = \nu_Q}} \widehat{\tau}_P^Q([\nu_P']^Q) t^{\langle \epsilon_P^Q, [\nu_P']^Q \rangle} = \Big(\prod_{\alpha \in \Delta_P^Q} \frac{1}{1 - t^{\langle \epsilon_P^Q, \alpha^\vee \rangle}} \Big) t^{\sum_{\alpha \in \Delta_P^Q} \langle \epsilon_P^Q, \alpha^\vee \rangle \langle \varpi_{\tilde{\alpha}}(\nu_Q) \rangle},$$

where, for each $\mu \in \mathbb{R}/\mathbb{Z}$, $\langle \mu \rangle \in \mathbb{R}$ is the unique representative of the class μ such that $0 < \langle \mu \rangle \leq 1$.

Notice that $< \cdot, \cdot >$ denotes the pairing between dual spaces, while $< \cdot >$ denotes the unique representative in $(0, 1]$ of the class $\cdot \in \mathbb{R}/\mathbb{Z}$.

PROOF. Given

$$\nu_Q \in \Lambda_{P_0}^Q = X_*(A'_{P_0}) \Big/ \bigoplus_{\alpha \in \Delta_{P_0}^Q} \mathbb{Z}\alpha^\vee,$$

we choose a representative $X_0 \in X_*(A'_{P_0}) \subset \mathfrak{a}_0$ of ν_Q. Let

$$\tilde{S} = \Big\{ X_0 + \sum_{\alpha \in \Delta_P^Q} m_\alpha \tilde{\alpha}^\vee \,\Big|\, m_\alpha \in \mathbb{Z} \Big\} \subset X_*(A'_{P_0}).$$

Then the natural projection

$$X_*(A'_{P_0}) \to \Lambda_{P_0}^P = X_*(A'_{P_0}) \Big/ \bigoplus_{\alpha \in \Delta_{P_0}^P} \mathbb{Z}\alpha^\vee$$

restricts to a bijection

$$j : \tilde{S} \xrightarrow{\cong} S = \{\nu_P \in \Lambda_{P_0}^P \mid [\nu_P]_Q = \nu_Q\}.$$

Let

$$f(t) = \sum_{\substack{\nu_P \in \Lambda_{P_0}^P \\ [\nu_P]_Q = \nu_Q}} \widehat{\tau}_P^Q([\nu_P']^Q) t^{\langle \epsilon_P^Q, [\nu_P']^Q \rangle}.$$

Then

$$f(t) = \sum_{\nu_P \in S} \widehat{\tau}_P^Q([\nu_P']^Q) t^{\langle \epsilon_P^Q, [\nu_P']^Q \rangle} = \sum_{\nu_P \in S_+} t^{\langle \epsilon_P^Q, [\nu_P']^Q \rangle},$$

where

$$S_+ = \{\nu_P \in S \mid \langle \varpi_\alpha^Q, [\nu_P']^Q \rangle > 0 \ \forall \alpha \in \Delta_P^Q\}.$$

Let $\tilde{S}_+ = j^{-1}(S_+)$. Then

$$\tilde{S}_+ = \Big\{ X_0 + \sum_{\alpha \in \Delta_P^Q} m_\alpha \tilde{\alpha}^\vee \,\Big|\, m_\alpha \in \mathbb{Z}, \varpi_{\tilde{\alpha}}(X_0) + m_\alpha > 0 \ \forall \alpha \in \Delta_P^Q \Big\}.$$

So

$$\begin{aligned} f(t) &= \sum_{\nu_P \in S_+} t^{\langle \epsilon_P^Q, [\nu'_P]^Q \rangle} = \sum_{X \in \tilde{S}_+} t^{\langle \epsilon_P^Q, [X]_P^Q \rangle} = \sum_{X \in \tilde{S}_+} \prod_{\alpha \in \Delta_P^Q} t^{\langle \epsilon_P^Q, \alpha^\vee \rangle \langle \varpi_\alpha^Q, [X]_P^Q \rangle} \\ &= \prod_{\alpha \in \Delta_P^Q} \sum_{\substack{m_\alpha \in \mathbb{Z} \\ \varpi_{\tilde{\alpha}}(X_0) + m_\alpha > 0}} t^{\langle \epsilon_P^Q, \alpha^\vee \rangle (\varpi_{\tilde{\alpha}}(X_0) + m_\alpha)} \end{aligned}$$

Note that $\langle \varpi_{\tilde{\alpha}}, \nu_Q \rangle = \varpi_{\tilde{\alpha}}(\nu_Q) = \varpi_{\tilde{\alpha}}(X_0) + \mathbb{Z} \in \mathbb{Q}/\mathbb{Z}$ for all $\alpha \in \Delta_P^Q$. As in the proof of [**LR**, Lemma 2.3], for $p \in \mathbb{Z}_{>0}$ and $x \in \mathbb{R}$, we have

$$\sum_{\substack{m \in \mathbb{Z} \\ x+m>0}} t^{p(x+m)} = \frac{t^{p\langle x+\mathbb{Z} \rangle}}{1 - t^p}.$$

Thus,

$$f(t) = \prod_{\alpha \in \Delta_P^Q} \frac{t^{\langle \epsilon_P^Q, \alpha^\vee \rangle \langle \varpi_{\tilde{\alpha}}(\nu_Q) \rangle}}{1 - t^{\langle \epsilon_P^Q, \alpha^\vee \rangle}}$$

□

Set $m(P, \nu'_P) = n_P + \langle \epsilon_P^G, \nu'_P \rangle$. We have now concluded with the following inversion formula, which is a slightly modified version of [**LR**, Theorem 2.4].

THEOREM A.6. *Given $a_0 : \mathcal{P} \to A$, there exists a unique function $b_0 : \mathfrak{T} \to A$ which satisfies the relation*

$$a_0(Q) = \sum_{\substack{P \in \mathcal{P} \\ P \subset Q}} \sum_{\substack{\nu_P \in \Lambda_{P_0}^P \\ [\nu_P]_Q = \nu_Q}} \tau_P^Q([\nu'_P]^Q) b_0(P, \nu_P) t^{m(P,\nu'_P) - m(Q,\nu'_Q)},$$

for each $(Q, \nu_Q) \in \mathfrak{T}$. This function is given by

$$b_0(Q, \nu_Q) =$$

$$\sum_{\substack{P \in \mathcal{P} \\ P \subset Q}} (-1)^{\dim(\mathfrak{a}_P^Q)} a_0(P) t^{n_P - n_Q} \Big(\prod_{\alpha \in \Delta_P^Q} \frac{1}{1 - t^{\langle \epsilon_P^Q, \alpha^\vee \rangle}} \Big) \cdot t^{\sum_{\alpha \in \Delta_P^Q} \langle \epsilon_P^Q, \alpha^\vee \rangle \langle \varpi_{\tilde{\alpha}}(\nu_Q) \rangle} \in A,$$

for each $(Q, \nu_Q) \in \mathfrak{T}$.

A.3. Inversion of the Atiyah-Bott recursion relation

Let $\mathcal{C}(G, \nu_G)$ be the space of complex structures on a C^∞ principal G-bundle over a Riemann surface of genus $g \geq 2$ with topological type $\nu_G \in \Lambda_{P_0}^G \cong \pi_1(G)$. Let $\mathcal{C}^{ss}(G, \nu_G) \subset \mathcal{C}(G, \nu_G)$ be the semi-stable stratum. Let $P_t(G, \nu_G)$ and $P_t^{ss}(G, \nu_G)$ be the $\mathcal{G}$-equivariant Poincaré series of $\mathcal{C}(G, \nu_G)$ and $\mathcal{C}^{ss}(G, \nu_G)$, respectively. Let $\mathcal{C}(G, P, \nu_P) \subset \mathcal{C}(G, \nu_G)$ be the stratum which corresponds to $(P, \nu_P) \in \mathfrak{T}$, where $[\nu_P]_G = \nu_G$. Then the real codimension $m(P, \nu'_P)$ of the stratum $\mathcal{C}(G, P, \nu_P)$ is equal to

$$2 \dim(N_P)(g-1) + 4\langle \rho_P^G, \nu'_P \rangle,$$

where N_P is the unipotent radical of P and

$$\rho_P^G = \frac{1}{2} \sum_{\alpha \in \Phi_P^{G+}} \alpha \in \mathfrak{a}_P^{G*} \subset \mathfrak{a}_P^*.$$

Clearly $m(G, \nu'_G) = 0$.

With the above notation, the Atiyah-Bott recursion relation can be stated as follows:

THEOREM A.7 (Atiyah-Bott). *The stratification of $\mathcal{M}(G,\nu_G)$ by the $\mathcal{M}(G,P,\nu_P)$ is perfect modulo torsion, so that for the Poincaré series, we have*

$$P_t(G,\nu_G) = \sum_{P\in\mathcal{P}} \sum_{\substack{\nu_P\in\Lambda^P_{P_0}\\ [\nu_P]_G=\nu_G}} \tau^G_P([\nu'_P]^G) t^{m(P,\nu'_P)} P^{ss}_t(M_P,\nu_P). \tag{A.3}$$

Note that Theorem A.7 and [**LR**, Theorem 3.2] are slightly different when G_{ss} is not simply connected.

THEOREM A.8 ([**LR**, Theorem 3.3]). *For any $\nu_G \in \Lambda^G_{P_0}$, we have*

$$P_t(G,\nu_G) = \Big(\frac{(1+t)^{2g}}{1-t^2}\Big)^{\dim(\mathfrak{a}_G)} \prod_{i=1}^{\dim(\mathfrak{a}^G_0)} \frac{(1+t^{2d_i(G)-1})^{2g}}{(1-t^{2d_i(G)-2})(1-t^{2d_i(G)})}\,.$$

In particular, $P_t(G,\nu_G)$ does not depend on ν_G.

Note that in both Theorem A.7 and Theorem A.8, we may replace G by the Levi component M_P of a parabolic subgroup P.

To invert the recursion relation (A.3), we apply Theorem A.6, with $a_0(P) = P_t(M_P,\nu_P)$, $b_0(P,\nu_P) = P^{ss}_t(M_P,\nu_P)$, $n_P = 2\dim(N_P)(g-1)$, $\epsilon^G_P = 4\rho^G_P$. We obtain

THEOREM A.9. *For any $\nu_G \in \Lambda^G_{P_0}$, we have*

$$P^{ss}_t(G,\nu_G) =$$

$$\sum_{P\in\mathcal{P}} (-1)^{\dim(\mathfrak{a}^G_P)} \Big(\frac{(1+t)^{2g}}{1-t^2}\Big)^{\dim(\mathfrak{a}_P)} \Big(\prod_{i=1}^{\dim(\mathfrak{a}^P_0)} \frac{(1+t^{2d_i(M_P)-1})^{2g}}{(1-t^{2d_i(M_P)-2})(1-t^{2d_i(M_P)})}\Big)$$

$$\cdot t^{2\dim(N_P)(g-1)} \Big(\prod_{\alpha\in\Delta^G_P} \frac{1}{1-t^{4\langle\rho^G_P,\alpha^\vee\rangle}}\Big) \cdot t^{4\sum_{\alpha\in\Delta^G_P}\langle\rho^G_P,\alpha^\vee\rangle\langle\varpi_{\tilde\alpha}(\nu_G)\rangle} \in \mathbb{Q}(t).$$

This is exactly Theorem 4.4.

Bibliography

[AB] M.F. Atiyah, R. Bott, *Yang-Mills equations over Riemann surfaces*, Philos. Trans. Roy. Soc. London Ser. A **308** (1983), no. 1505, 523–615. MR702806 (85k:14006)

[AuB] D.M. Austin, P.J. Braam, *Morse-Bott theory and equivariant cohomology*, The Floer memorial volume, 123–183, Progr. Math., 133, Birkhäuser, Basel, 1995. MR1362827 (96i:57037)

[B] T. Baird, *The moduli space of flat SU(2)-connections over a nonorientable surface*, Ph.D. thesis, University of Toronto, 2007.

[BD] T. Bröcker, T. tom Dieck, *Representations of compact Lie groups*, Graduate Texts in Mathematics, **98**. Springer-Verlag, New York, 1985. MR781344 (86i:22023)

[BGH] D. Biss, V. Guillemin, T.S. Holm, *The mod 2 cohomology of fixed point sets of anti-symplectic involutions*, Adv. Math. **185** (2004), no. 2, 370–399. MR2060474 (2005c:53104)

[Da] Daskalopoulos, *The topology of the space of stable bundles on a compact Riemann surface*, J. Differential Geometry (1992), no. 36, 699-746. MR1189501 (93i:58026)

[FH] W. Fulton, J. Harris, *Representation theory, a first course*, Graduate Texts in Mathematics, **129**. Readings in Mathematics, Springer-Verlag, New York, 1991. MR1153249 (93a:20069)

[FM] R. Friedman, J. Morgan, *On the converse to a theorem of Atiyah and Bott*, J. Algebraic Geom. **11** (2002), no. 2, 257–292. MR1874115 (2003b:14041)

[GH] R.F. Goldin, T.S. Holm, *Real loci of symplectic reductions*, Trans. Amer. Math. Soc. **356** (2004), no. 11, 4623–4642. MR2067136 (2005e:53135)

[Ho] N.-K. Ho, *The real locus of an involution map on the moduli space of flat connections on a Riemann surface*, Int. Math. Res. Not. (2004), no. 61, 3263–3285. MR2096257 (2005g:53167)

[HL1] N.-K. Ho, C.-C.M. Liu, *On the connectedness of moduli spaces of flat connections over compact surfaces*, Canad. J. Math. **56** (2004), no. 6, 1228–1236. MR2102631 (2005h:53148)

[HL2] N.-K. Ho, C.-C.M. Liu, *Connected components of the space of surface group representations*, Int. Math. Res. Not. (2003), no. 44, 2359–2372. MR2003827 (2004h:53116)

[HL3] N.-K. Ho, C.-C.M. Liu, *Connected components of the space of surface group representations II*, Int. Math. Res. Not. (2005), no. 16, 959–979. MR2146187 (2006b:53108)

[HL4] N.-K. Ho, C.-C.M. Liu, *Yang-Mills connections on nonorientable surfaces*, Comm. Anal. Geom. **16** (2008), no. 3, 617–679. MR2429971

[HuL] D. Huybrechts, M. Lehn, *The Geometry of moduli spaces of sheaves*, Aspects of Mathematics, E31. Friedr. Vieweg & Sohn, Braunschweig, 1997. MR1450870 (98g:14012)

[HN] G. Harder, M.S. Narasimhan, *On the cohomology groups of moduli spaces of vector bundles on curves*, Math. Ann. 212 (1975), 215–248. MR0364254 (51:509)

[Ki] Y.-H. Kiem, *Intersection cohomology of representation spaces of surface groups*, Int. J. Math. **17** (2006), no. 2, 169–182. MR2205432 (2007b:53174)

[K1] F.Kirwan, *Cohomology of quotients in symplectic and algebraic geometry*, Mathematical Notes, 31. Princeton University Press, Princeton, NJ, 1984. MR766741 (86i:58050)

[K2] F.Kirwan, *On spaces of maps from Riemann surfaces to Grassmannians and applications to the cohomology of moduli of vector bundles*, Ark. Mat. **24** (1986), 221–275. MR884188 (88h:14014)

[K3] F.Kirwan, *On the homology of compactifications of moduli spaces of vector bundles over a Riemann surface*, Proc. London Math. Soc. **53** (1986), 237–266. MR850220 (88e:14012)

[K4] F.Kirwan, *The cohomology rings of moduli spaces of bundles over Riemann surfaces*, J. Amer. Math. Soc. **5** (1992), 853–906. MR1145826 (93g:14016)

[Kn] A. Knapp, *Lie groups beyond an introduction*, Progress in Mathematics, **140**. Birkhäuser Boston, Inc., Boston, MA, 1996. MR1399083 (98b:22002)

[L] E. Lerman, *Gradient flow of the norm squared of a moment map*, Enseign. Math. (2) **51** (2005), no. 1-2, 117–127. MR2154623 (2006b:53106)

[LR] G. Laumon, M. Rapoport, *The Langlands lemma and the betti numbers of stacks of G-bundle on a curve*, Internat. J. Math. **7** (1996), no. 1, 29–45. MR1369904 (97f:14012)

[LM] H.B. Lawson, M.-L. Michelsohn, *Spin geometry*, Princeton Mathematical Series, **38**. Princeton University Press, Princeton, NJ, 1989. MR1031992 (91g:53001)

[M] J.W. Milnor, J.D. Stasheff, *Characteristic classes*, Annals of Mathematics Studies, **76**. Princeton University Press, Princeton, NJ, University of Tokyo Press, Tokyo, 1974. MR0440554 (55:13428)

[MFK] D. Mumford, J. Fogarty, F. Kirwan, *Geometric invariant theory*, Third edition. Ergebnisse der Mathematik und ihrer Grenzgebiete (2) [Results in Mathematics and Related Areas (2)], **34**. Springer-Verlag, Berlin, 1994. MR1304906 (95m:14012)

[Mu] J.R. Munkres, *Elements of algebraic topology*, Addison-Wesley Publishing Company, Menlo Park, CA, 1984. MR755006 (85m:55001)

[Rå] J. Råde, *On the Yang-Mills heat equation in two and three dimensions*, J. Reine Angew. Math **431** (1992), 123–163. MR1179335 (94a:58041)

[Ra] A. Ramanathan, *Stable Principal Bundles on a Compact Riemann Surface*, Math. Ann. **213** (1975), 129–152. MR0369747 (51:5979)

[Se] J.-P. Serre, *Groupes d'homotopie et classes de groupes abéliens*, Ann. of Math. **58** (1953), 258–294. MR0059548 (15:548c)

[UY] K. Uhlenbeck, S.-T. Yau, *On the existence of Hermitian-Yang-Mills connections in stable vector bundles*, Frontiers of the mathematical sciences: 1985 (New York, 1985). Comm. Pure Appl. Math. **39** (1986), no. S, suppl., S257–S293. MR861491 (88i:58154)

[Wa] S. Wang, *A Narasimhan-Seshadri-Donaldson correspondence over non-orientable surfaces*, Forum Math. **8** (1996), no. 4, 461–474. MR1393324 (98h:53043)

[Wo] C. Woodward, *The Yang-Mills heat flow on the moduli space of framed bundles on a surface*, Amer. J. Math. **128** (2006), no. 2, 311-359. MR2214895 (2007d:53114)

[Za] D. Zagier, *Elementary aspects of the Verlinde formula and of the Harder-Narasimhan-Atiyah-Bott formula*, Proceedings of the Hirzebruch 65 Conference on Algebraic Geometry (Ramat Gan, 1993), 445–462, Israel Math. Conf. Proc., **9**, Bar-Ilan Univ., Ramat Gan, 1996. MR1360519 (96k:14005)

Editorial Information

To be published in the *Memoirs*, a paper must be correct, new, nontrivial, and significant. Further, it must be well written and of interest to a substantial number of mathematicians. Piecemeal results, such as an inconclusive step toward an unproved major theorem or a minor variation on a known result, are in general not acceptable for publication.

Papers appearing in *Memoirs* are generally at least 80 and not more than 200 published pages in length. Papers less than 80 or more than 200 published pages require the approval of the Managing Editor of the Transactions/Memoirs Editorial Board. Published pages are the same size as those generated in the style files provided for AMS-LaTeX or AMS-TeX.

Information on the backlog for this journal can be found on the AMS website starting from `http://www.ams.org/memo`.

A Consent to Publish and Copyright Agreement is required before a paper will be published in the *Memoirs*. After a paper is accepted for publication, the Providence office will send a Consent to Publish and Copyright Agreement to all authors of the paper. By submitting a paper to the *Memoirs*, authors certify that the results have not been submitted to nor are they under consideration for publication by another journal, conference proceedings, or similar publication.

Information for Authors

Memoirs is an author-prepared publication. Once formatted for print and on-line publication, articles will be published as is with the addition of AMS-prepared frontmatter and backmatter. Articles are not copyedited; however, confirmation copy will be sent to the authors.

Initial submission. The AMS uses Centralized Manuscript Processing for initial submissions. Authors should submit a PDF file using the Initial Manuscript Submission form found at `www.ams.org/peer-review-submission`, or send one copy of the manuscript to the following address: Centralized Manuscript Processing, MEMOIRS OF THE AMS, 201 Charles Street, Providence, RI 02904-2294 USA. If a paper copy is being forwarded to the AMS, indicate that it is for *Memoirs* and include the name of the corresponding author, contact information such as email address or mailing address, and the name of an appropriate Editor to review the paper (see the list of Editors below).

The paper must contain a *descriptive title* and an *abstract* that summarizes the article in language suitable for workers in the general field (algebra, analysis, etc.). The *descriptive title* should be short, but informative; useless or vague phrases such as "some remarks about" or "concerning" should be avoided. The *abstract* should be at least one complete sentence, and at most 300 words. Included with the footnotes to the paper should be the 2010 *Mathematics Subject Classification* representing the primary and secondary subjects of the article. The classifications are accessible from `www.ams.org/msc/`. The Mathematics Subject Classification footnote may be followed by a list of *key words and phrases* describing the subject matter of the article and taken from it. Journal abbreviations used in bibliographies are listed in the latest *Mathematical Reviews* annual index. The series abbreviations are also accessible from `www.ams.org/msnhtml/serials.pdf`. To help in preparing and verifying references, the AMS offers MR Lookup, a Reference Tool for Linking, at `www.ams.org/mrlookup/`.

Electronically prepared manuscripts. The AMS encourages electronically prepared manuscripts, with a strong preference for AMS-LaTeX. To this end, the Society has prepared AMS-LaTeX author packages for each AMS publication. Author packages include instructions for preparing electronic manuscripts, samples, and a style file that generates the particular design specifications of that publication series. Though AMS-LaTeX is the highly preferred format of TeX, author packages are also available in AMS-TeX.

Authors may retrieve an author package for *Memoirs of the AMS* from `www.ams.org/journals/memo/memoauthorpac.html` or via FTP to `ftp.ams.org` (login as `anonymous`, enter your complete email address as password, and type `cd pub/author-info`). The

AMS Author Handbook and the *Instruction Manual* are available in PDF format from the author package link. The author package can also be obtained free of charge by sending email to tech-support@ams.org (Internet) or from the Publication Division, American Mathematical Society, 201 Charles St., Providence, RI 02904-2294, USA. When requesting an author package, please specify $\mathcal{A}\mathcal{M}\mathcal{S}$-LaTeX or $\mathcal{A}\mathcal{M}\mathcal{S}$-TeX and the publication in which your paper will appear. Please be sure to include your complete mailing address.

After acceptance. The source files for the final version of the electronic manuscript should be sent to the Providence office immediately after the paper has been accepted for publication. The author should also submit a PDF of the final version of the paper to the editor, who will forward a copy to the Providence office.

Accepted electronically prepared files can be submitted via the web at www.ams.org/submit-book-journal/, sent via FTP, or sent on CD-Rom or diskette to the Electronic Prepress Department, American Mathematical Society, 201 Charles Street, Providence, RI 02904-2294 USA. TeX source files and graphic files can be transferred over the Internet by FTP to the Internet node ftp.ams.org (130.44.1.100). When sending a manuscript electronically via CD-Rom or diskette, please be sure to include a message indicating that the paper is for the *Memoirs*.

Electronic graphics. Comprehensive instructions on preparing graphics are available at www.ams.org/authors/journals.html. A few of the major requirements are given here.

Submit files for graphics as EPS (Encapsulated PostScript) files. This includes graphics originated via a graphics application as well as scanned photographs or other computer-generated images. If this is not possible, TIFF files are acceptable as long as they can be opened in Adobe Photoshop or Illustrator.

Authors using graphics packages for the creation of electronic art should also avoid the use of any lines thinner than 0.5 points in width. Many graphics packages allow the user to specify a "hairline" for a very thin line. Hairlines often look acceptable when proofed on a typical laser printer. However, when produced on a high-resolution laser imagesetter, hairlines become nearly invisible and will be lost entirely in the final printing process.

Screens should be set to values between 15% and 85%. Screens which fall outside of this range are too light or too dark to print correctly. Variations of screens within a graphic should be no less than 10%.

Inquiries. Any inquiries concerning a paper that has been accepted for publication should be sent to memo-query@ams.org or directly to the Electronic Prepress Department, American Mathematical Society, 201 Charles St., Providence, RI 02904-2294 USA.

Editors

This journal is designed particularly for long research papers, normally at least 80 pages in length, and groups of cognate papers in pure and applied mathematics. Papers intended for publication in the *Memoirs* should be addressed to one of the following editors. The AMS uses Centralized Manuscript Processing for initial submissions to AMS journals. Authors should follow instructions listed on the Initial Submission page found at `www.ams.org/memo/memosubmit.html`.

Algebra, to ALEXANDER KLESHCHEV, Department of Mathematics, University of Oregon, Eugene, OR 97403-1222; e-mail: ams@noether.uoregon.edu

Algebraic geometry, to DAN ABRAMOVICH, Department of Mathematics, Brown University, Box 1917, Providence, RI 02912; e-mail: amsedit@math.brown.edu

Algebraic geometry and its applications, to MINA TEICHER, Emmy Noether Research Institute for Mathematics, Bar-Ilan University, Ramat-Gan 52900, Israel; e-mail: teicher@macs.biu.ac.il

Algebraic topology, to ALEJANDRO ADEM, Department of Mathematics, University of British Columbia, Room 121, 1984 Mathematics Road, Vancouver, British Columbia, Canada V6T 1Z2; e-mail: adem@math.ubc.ca

Combinatorics, to JOHN R. STEMBRIDGE, Department of Mathematics, University of Michigan, Ann Arbor, Michigan 48109-1109; e-mail: JRS@umich.edu

Commutative and homological algebra, to LUCHEZAR L. AVRAMOV, Department of Mathematics, University of Nebraska, Lincoln, NE 68588-0130; e-mail: avramov@math.unl.edu

Complex analysis and harmonic analysis, to ALEXANDER NAGEL, Department of Mathematics, University of Wisconsin, 480 Lincoln Drive, Madison, WI 53706-1313; e-mail: nagel@math.wisc.edu

Differential geometry and global analysis, to CHRIS WOODWARD, Department of Mathematics, Rutgers University, 110 Frelinghuysen Road, Piscataway, NJ 08854; e-mail: ctw@math.rutgers.edu

Dynamical systems and ergodic theory and complex analysis, to YUNPING JIANG, Department of Mathematics, CUNY Queens College and Graduate Center, 65-30 Kissena Blvd., Flushing, NY 11367; e-mail: Yunping.Jiang@qc.cuny.edu

Functional analysis and operator algebras, to DIMITRI SHLYAKHTENKO, Department of Mathematics, University of California, Los Angeles, CA 90095; e-mail: shlyakht@math.ucla.edu

Geometric analysis, to WILLIAM P. MINICOZZI II, Department of Mathematics, Johns Hopkins University, 3400 N. Charles St., Baltimore, MD 21218; e-mail: trans@math.jhu.edu

Geometric topology, to MARK FEIGHN, Math Department, Rutgers University, Newark, NJ 07102; e-mail: feighn@andromeda.rutgers.edu

Harmonic analysis, representation theory, and Lie theory, to ROBERT J. STANTON, Department of Mathematics, The Ohio State University, 231 West 18th Avenue, Columbus, OH 43210-1174; e-mail: stanton@math.ohio-state.edu

Logic, to STEFFEN LEMPP, Department of Mathematics, University of Wisconsin, 480 Lincoln Drive, Madison, Wisconsin 53706-1388; e-mail: lempp@math.wisc.edu

Number theory, to JONATHAN ROGAWSKI, Department of Mathematics, University of California, Los Angeles, CA 90095; e-mail: jonr@math.ucla.edu

Number theory, to SHANKAR SEN, Department of Mathematics, 505 Malott Hall, Cornell University, Ithaca, NY 14853; e-mail: ss70@cornell.edu

Partial differential equations, to GUSTAVO PONCE, Department of Mathematics, South Hall, Room 6607, University of California, Santa Barbara, CA 93106; e-mail: ponce@math.ucsb.edu

Partial differential equations and dynamical systems, to PETER POLACIK, School of Mathematics, University of Minnesota, Minneapolis, MN 55455; e-mail: polacik@math.umn.edu

Probability and statistics, to RICHARD BASS, Department of Mathematics, University of Connecticut, Storrs, CT 06269-3009; e-mail: bass@math.uconn.edu

Real analysis and partial differential equations, to DANIEL TATARU, Department of Mathematics, University of California, Berkeley, Berkeley, CA 94720; e-mail: tataru@math.berkeley.edu

All other communications to the editors, should be addressed to the Managing Editor, ROBERT GURALNICK, Department of Mathematics, University of Southern California, Los Angeles, CA 90089-1113; e-mail: guralnic@math.usc.edu.

Titles in This Series

For a complete list of titles in this series, visit the AMS Bookstore at **www.ams.org/bookstore/**.